The **Beauty** of **Elementary Mathematics**

And How to Teach It

The **Beauty** of **Elementary Mathematics**

And How to Teach It

Ron Aharoni

Technion, Israel Institute of Technology,
Haifa, Israel

World Scientific

NEW JERSEY · LONDON · SINGAPORE · BEIJING · SHANGHAI · HONG KONG · TAIPEI · CHENNAI · TOKYO

Published by

World Scientific Publishing Co. Pte. Ltd.

5 Toh Tuck Link, Singapore 596224

USA office: 27 Warren Street, Suite 401-402, Hackensack, NJ 07601

UK office: 57 Shelton Street, Covent Garden, London WC2H 9HE

Library of Congress Cataloging-in-Publication Data

Names: Aharoni, Ron, author.

Title: The beauty of elementary mathematics : and how to teach it /
Ron Aharoni, Technion, Israel Institute of Technology, Haifa, Israel.

Other titles: Beauty of elementary mathematics and how to teach it

Description: Singapore ; Hackensack, NJ : World Scientific, [2024] | Includes index.

Identifiers: LCCN 2024018786 | ISBN 9789811278150 (hardcover) |
ISBN 9789811278747 (paperback) | ISBN 9789811278167 (ebook for institutions) |
ISBN 9789811278174 (ebook for individuals)

Subjects: LCSH: Mathematics--Study and teaching (Elementary)

Classification: LCC QA135.6 .A373 2024 | DDC 372.7--dc23/eng/20240509

LC record available at https://lccn.loc.gov/2024018786

British Library Cataloguing-in-Publication Data

A catalogue record for this book is available from the British Library.

Cover image: "Red Balloon" (1922) by Paul Klee (1879–1940).

Collection/Source/Photographer - Solomon R. Guggenheim Museum

References - Solomon R. Guggenheim Foundation Artwork ID: 2143

For any available supplementary material, please visit
https://www.worldscientific.com/worldscibooks/10.1142/13465#t=suppl

Desk Editors: Nambirajan Karuppiah/Rok Ting Tan

Typeset by Stallion Press
Email: enquiries@stallionpress.com

About the Author

Ron Aharoni is a retired professor of mathematics. Most of his career he spent at the mathematics department at the Technion, Israel Institute of Technology. He spent ten years tutoring teachers and teaching in elementary schools. He was active in the struggles that shaped the mathematical education in Israel. He is the author of seven books, translated into many languages. Among them—*Mathematics, Poetry and Beauty, Man Detaches Meaning*—a book on techniques common to poetry and jokes.

Acknowledgments

All I know about elementary school education I owe to teachers, who were patient enough with the vanity of a math professor. In particular, the infinite wisdom of Talma Gavish, the ultimate teacher, who gave me for free the fruit of her lifelong experience. I am grateful to Hanna Heller, who read and commented, and of course to my first and last editor, my wife Merav.

Last, but definitely not least, my gratitude goes to the illustrator, Ayelet Bar-Noy, who delved into an unfamiliar field, learned it and internalized its spirit.

Contents

Introduction

Some 20 years ago, I took a partial leave from my university position to instruct elementary school teachers. One day when I returned home, my 7-year-old daughter Inbal asked me what I had taught. "Why 2 times 3 is 3 times 2," I answered. "But that's the same thing!" she remonstrated. I asked her to raise 3 fingers in each hand — that's 2 times 3. "Now raise 3 times 2 fingers." She raised two pairs of fingers, and I provided the third pair with my fingers. "Now show me they are equal," I said. To my amazement, she raised 3 fingers in each hand (2 times 3), pressed the two triples against each other, and spread the three pairs — 3 times 2.

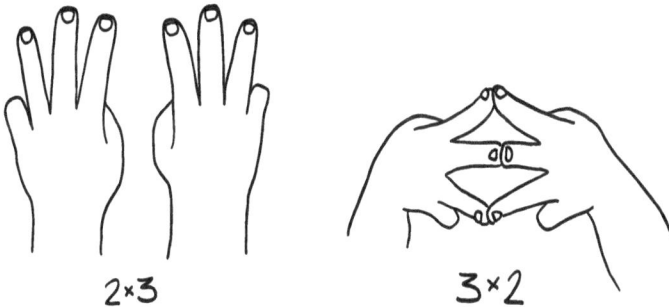

2 times 3 is 3 times 2

The exchangeability of the multiplied numbers is called *the commutativity law of multiplication*. The reader is encouraged to use this trick to show that 2 times 5 is 5 times 2.

A couple of years later, in another school, I showed this trick in Grade 2. A student in the first row mumbled — "that's beautiful."

My thoughts sometimes return to him. How often did he experience the beauty of math? Did it make him love the subject? I would like to believe so, but it is also possible that, like most of his friends, he wrestled the beast of mathematics throughout his school years. Why is the most beautiful of all school subjects such a source of agony?

"There are no learning disabilities," goes a common saying, "only teaching disabilities." If only teachers knew how to teach, math lessons would be bliss. True? Probably not. Math has always been the hardest school subject and will probably remain so. "There is no king's road to math," the Greek mathematician Euclid told Ptolemy, the King of Alexandria who consulted with him on how to learn geometry. Abstraction power is not distributed evenly among human beings. Some receive it as a gift from birth, but most others must struggle to acquire it. In history classes, you meet conflicts and passions familiar from everyday life. Mathematical concepts are more remote.

Many magic remedies have been prescribed. The Dutch tried teaching arithmetic (the theory of numbers) through money. Other countries used applications to everyday life. In the US of the 1960s and 1970s, an approach was attempted to start from the "top," namely abstractions first. Georges Cuisenaire, a Belgian teacher, invented rods of different lengths and colors to represent numbers concretely (in Israel, they even had a "zero" rod that does not exist but is colored pink). Some 30 years ago, a method of "investigation" was tried, in which students discovered mathematical principles on their own, by experimentation.

All to no avail. The countries leading in the international exams (nowadays Finland and Far East countries) use the conservative approach, the old step-by-step methods. Indeed, this is the secret: the only way is step by step. How to do it, and how to impress the beauty of the mathematical edifice on the students, is the topic of this book.

What's in the Book

A previous book of mine, *Arithmetic for Parents*, published 20 years ago, dealt with one side of the elementary school curriculum: arithmetic, namely numbers. The present book also includes the other part of the curriculum: geometry. It also peeks at kindergarten: how to teach basic thought principles to preschool children. There is also a bit about middle school. There is more stress on the principles of teaching and thinking. One can also find some echo of a change in my own attitude, a byproduct of age: a shift of weight from math to life.

In geometry, the book describes how to imbibe the principles through experimentation. Familiarization with the concepts through drawing and observation. We shall follow in the footsteps of the Greeks in their admiration of the beauty of the subject. Children and grownups alike can appreciate this beauty. To get to the gems of this area, I took the liberty of introducing some more advanced topics.

Special stress is put on the circle, the most fascinating shape in elementary geometry. We shall learn three advanced topics that necessitate a rudimentary (Greek-version) knowledge of calculus: (1) the area of the circle, (2) the area of the sphere, and (3) the volume of the ball.

There are two types of chapters: (1) about mathematical principles and (2) on teaching and thinking principles. The headings of the former are marked by the 📖 icon, and the headings of the latter are marked by the 🖎 icon. There is also an account of the fascinating politics of education. In particular, the ease with which educational revolutions take over entire systems, and how, peculiarly, they invariably go against basic teaching principles.

When writing the book, I had three types of readers in mind. The first audience are people who wish to grapple with their old demons to better understand their childhood studies, the second are teachers, and the third are parents. Why the third one? Parents are the principal mediators of thinking. The child adopts and imitates their modes of thinking in mathematics as in life. Mathematical literacy of parents passes on to their children, if not by direct teaching, then by diffusion.

Part 1
Thinking Mathematics

Beauty

Beauty is in the blind spot of the beholder.

— Anonymous

The most beautiful things in the world cannot be seen or touched, they are felt with the heart.

— *The Little Prince*, Antoine de Saint-Exupéry

Ask a mathematician what attracts him or her to their profession, and in nine cases out of ten, the answer will be "beauty." Edna St. Vincent Millay (1892–1950), an American poet, wrote "Euclid alone has looked on beauty bare." The mathematician G.H. Hardy (1877–1947) said "There is no place under the sky for ugly mathematics" and claimed that Archimedes (the great Sicilian mathematician whose name will emerge in the book again and again) will be remembered long after Homer is forgotten.

What makes mathematics beautiful? Here, I should have opened a book-length digression to address the time-old question: "what is beauty?" I did this in another book, *Mathematics, Poetry and Beauty* (World Scientific, 2018). Here, I mention only two principles:

1. We sense beauty when either our emotions or our intellect are touched.
2. The sensation involved should be only half-conscious. We are touched by a feather. Beauty is in the blind spot of the beholder.

Touching the intellect means discovering order. But for the order to be beautiful, it must be too deep and complex to be fully grasped. It should be surprising so that we do not have time to follow it in its entirety. As for emotions, we feel beauty when we are not sure what string in our mind was plucked. This is, for example, the secret of the poetic metaphor — its beauty is in the uncertainty — is it the metaphoric meaning or the face-value one? The message is conveyed on two levels, and we are not sure which one we perceive. Knowing and not knowing at the same time.

A necessary requirement for the "not knowing" side is condensation. So much is squeezed into so few words or symbols that you cannot grasp it all consciously. Nothing does it better than formulas. They are as concise as can be. Einstein's formula for the conversion of mass to energy

$$E = mc^2$$

contains a whole world — the entire special theory of relativity compressed into a few symbols. And look how beautiful the similarity is to the formula of the kinetic energy of a body:

$$E = \frac{1}{2}mv^2$$

Here, v is the velocity of the body. In both formulas, the energy is expressed via the mass times the square of some speed. There must be some connection — but what is it?

It is condensation that enables us to enjoy the same piece of art again and again. A Mozart symphony has so much going on in it, that even professional musicians will never exhaust its depths. Basho, a Haiku poet (1644–1694) said that if a poem hides 80% of itself, we shall never tire of it. The same is true in mathematics. A new concept or idea introduces surprising order, but you only see the tip of the iceberg, it contains more than can be immediately grasped.

Beauty goes hand in hand with truth. The Czech writer Kafka (1883–1924) said that poetry is always a search for the truth. So is mathematics. "If I have to choose between beauty and truth", said the mathematician and physicist Harmann Weyl (1885–1955), "I choose beauty".

What Is Mathematics?

Mathematicians may practice mathematics for its sense of beauty; they are paid for its effectivity. Eugene Wigner (1902–1995), a physics Nobel laureate, spoke of "the unreasonable efficiency of mathematics in the natural sciences." Steven Weinberg, another physics Nobel laureate, said that he was always surprised by the fertility of mathematics and the barrenness of philosophy (In another book, *The Cat That is Not There: A Non-Philosophical Book About Philosophy*, Magness Press, 2009, Hebrew, I attempt to explain the latter).

Paul Dirac (1902–1984), one of the fathers of quantum theory, claimed that God is a first-rate mathematician who, in creating the universe, applied (or may still be applying) the most advanced, up-to-date mathematical tools. Strange, isn't it? This would mean that in 50 years God will have even better tools and will construct the world differently!

What is the secret of mathematics? To answer, we must first understand what it is. I like to tease freshmen with this question. "Are you planning to spend four years of your life on something you cannot even define?" In fact, they are not to blame. Even professional mathematicians will falter. Most never even address the question. "A mathematician does not know what he is speaking about," said Bertrand Russel, mathematician in his youth, philosopher at old age (1872–1970). Still, the mathematician fares better than the philosopher, whom William James (1842–1910, a philosopher himself) described as a "blind person, looking in a dark room, for a black cat that is not there."

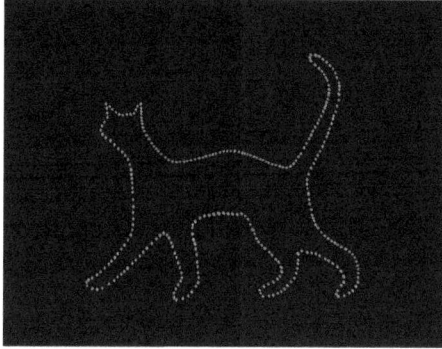

A layman's answer usually includes the word "numbers." Math is the science of numbers. Of course, this is off the mark — there is also geometry. "Mathematics is the science of numbers, shapes and order," is an attempt at rectification. Still vague — doesn't every science deal with order?

The following definition is more to the point:

Mathematics deals with extreme abstractions.

Mathematics abstracts the most basic mental processes. The first and foremost of those is discerning the objects around us: table, chair, apple. When an object repeats, we count: 2 apples, 3 apples, 4 apples. Later, we discover that the rules governing the numbers do not depend on the type — whether these are 3 apples or 3 pencils. So, we construct the abstract concept "3."

Mathematics strips the tree from its leaves ("apples, pencils") and looks at the trunk — the number of objects, no matter which. This saves energy: if joining 2 apples to 3 apples gives 5 apples, the same will be true for pencils. So, we write abstractly, $2 + 3 = 5$. "Math is for the lazy," said the Hungarian mathematician Georg Polya, "the principles work for you."

The same goes for geometry. If two lines on my page meet at most at a single point, this will be true for every pair of lines. It is a general rule. "Abstraction" is synonymous with "generality."

Mathematics and the Humanities

What is the difference between mathematics and philosophy? In mathematics someone important is someone who said something important. In philosophy something important is something that someone important said.

— Anonymous

I always wonder, why people value themselves more than others, but value others' opinions about themselves more than their own.

— Marcus Aurelius, Roman Emperor and Philosopher

A twin question to "what is mathematics" is: "how do mathematicians think?" What characterizes the mathematical way of thinking? Here are a few characteristics, with emphasis on the differences from the humanities:

- In the humanities, experience is an advantage. Mathematics is for the young. The Fields Medal, which is the mathematical analog of the Nobel Prize, is restricted to people younger than 40. In mathematics, most of the major discoveries were made by young people. Galois (1811–1832) laid the foundations of modern algebra in a will he wrote the night before he was killed in a pointless duel at the age of 20. Newton invented calculus at the age of 24. When he was 23 years old, Kurt Gödel proved a theorem that changed the course of mathematics — not every true fact about numbers can be proved from the standard axioms. A young brain can form new patterns.

- Math is democratic. In the sciences, "who said" is of no importance. In the humanities, it is often more important who said than what he or she said — breeding the strange phenomenon of experts on the writings of previous thinkers. In a math department you will not find an expert on Archimedes. In philosophy departments there are experts on Plato.
- In the humanities, knowledge grows like stalactites, in slow motion and gradually. A mathematical discovery arrives (following hard work and long hatching) in a flash illumination.
- In science, old results are quickly subsumed by new ones. The physics of ancient Greece is of interest to historians, not to physicists. Life wisdom, by contrast, is timeless: open an ancient parables book and you will find wisdom relevant to this day, such as the Marcus Aurelius quotation at the opening of this chapter.

Mathematics is constructed layer upon layer. The building blocks in mathematics are well polished, fitting each other and enabling the construction of tall buildings. In modern mathematics, the towers are sometimes so high that the help of a computer must be summoned.

Part 2
Didactics

Layer upon Layer

The layered structure of mathematics dictates the first duty of a math teacher: seeing to it that no rung is skipped. This is easier said than done, since in the brains of adults the rungs are not explicit, they are too deeply ingrained. It is easy to miss a layer, with the same result as that of missing a floor in a high-rise building.

Unfortunately, this happens all too often. Students are not taught the most basic facts. Do you remember having been asked any of the following questions:

- Why are arithmetic notebooks checkered?
- Why do you start the calculation of addition, subtraction, and multiplication from the right, and division from the left?
- What, exactly, is a fraction?
- Why is $\frac{1}{2}$ of 10 the same as $\frac{1}{2} \times 10$? (a question even professional mathematicians have a hard time answering)
- What does 3 times 4 mean?

And dozens of other questions that the students are not encouraged to ask. Often, they are discouraged. "Why" discussions are as rare as rain in the Kalahari desert. In middle school, the students are taught the formula for the solution of quadratic equations as a black box, with no explanation.

It is not even tested in simple equations like $x^2 = 0$. This is how math anxiety is born — from the unknown. "The concepts have not been solidly structured in my brain" becomes "I am stupid."

One aim of this book is to fill in the gaps. Turning our attention to these will catch two birds with one stone — getting acquainted with the beauty of elementary math, and learning how to teach it.

How Do You Catch a Cloud and Pin It Down?

How do you solve a problem like Maria? How do you hold a moonbeam in your hand?

— *The Sound of Music*, 1965

Carl Friedrich Gauss (1777–1855) was the greatest mathematician of the 19th century, some say of all time. When asked how he thinks, he answered: "methodically and concretely." "Methodically" means not skipping stages; "Concretely" means through examples. Every mathematician knows: the first step towards solving a problem is considering concrete special cases. The abstractions will arrive by themselves.

How do you pin down abstract concepts? — by starting with the concrete. Experienced teachers do not explain, they show. Up to Grade 3, and even beyond, the student's desk must be equipped with all kinds of counting objects: pebbles, beads, buttons. My favorites are popsicle sticks of the larger type, those used by doctors, since they can be shown to the entire class. "Mary, can you show us 4+3?" And the entire class follows Mary joining 4 sticks in her right hand with 3 sticks in her left.

In Grade 1, you should count, count, count. To avoid fixation, use different types of objects (remember? abstract = general). Then count backward — after positioning 20 beads on your desk, remove them one by one and count 19, 18, 17... I will later explain the importance of going backwards. When you get to the decimal system, collect 10 beads to a "ten," then 10 tens to a hundred, then 10 hundreds to a thousand — all concretely. Hands-on experience is never a waste of time.

All this is self-evident, isn't it? Not quite. In the 1950s, a new didactic approach took over in the United States, the "New Math" — starting from the abstract. "Our children will start from the point where scientists are," boasted the originators of the method. As in life, cutting corners never pays. In the US, the not-unexpected crash that ensued sent the education system back to normal, but foreign Ph.D. students who took their degrees in the US during that period brought the gospel ("plague" is more appropriate) to their countries to spread havoc there. In Israel, the folly lasted into the 21st century.

Three Stages: Concrete, Pictorial, Abstract

For many years, Singapore led the scores in the international exams in elementary schools. Nowadays, China is usually first. I am not familiar with the Chinese system, but I followed the Singapore system closely. On a visit there, I was thoroughly impressed by the serious attitude of the students. There were a lot of individual chats, but always on the topic of the lesson. There is tremendous motivation, in fact too much of it. On weekends, the children take private lessons from university professors. But above all, the system follows sound teaching principles.

One of these is constructing the concepts in three steps: concrete $\rightarrow$ pictorial $\rightarrow$ abstract. Hands on, then drawing, and finally putting the conclusions into words.

Take, for example, the calculation of addition. We start with the concrete — let Ana hold 8 popsicle sticks, Bob 5 sticks, and ask Bob to complete Ana's bundle to 10 — how many should he give? Indeed, 2. We are left with 10 (in Ana's hand) and 3 (in Bob's hand). Draw it on the board — 1 ten above Ana, 3 above Bob.

A second stage is replacing the sticks by pegs drawn on paper. Finally, discussion: how many sticks should Ana give, to complete a 10? Indeed, 5. What is easier? Ana is closer to 10, so it is more economical to complete her 10.

After several of these examples, you may move on to a question demanding more abstraction, like 999 + 8.

As another example of the three stages, take the decimal system, which is about collecting tens. We start with gathering objects into groups of 10: wrap strings around tens of popsicle sticks, thread tens of beads on strings, then collect ten tens and call the resulting group "a hundred."

The next stage is drawing. Draw a lot of lines (or flowers if you have the patience) and circle tens of them, then collect the circles in groups of 10, naming them "hundreds." More and more, until the students start laughing at the repetition.

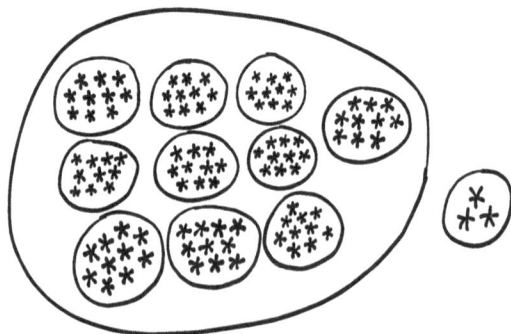

Then discussion. Why did we collect tens? (answer: it is easier to see how many we have). Why tens and not eights? (answer: we have ready-made 10 fingers, which we can count).

How many times is 10 larger than 1? How many tens go into a 100? How many times is 1000 larger than 100? Each step is 10 times its predecessor. There are plenty of points that can be discussed.

Pace

Why start with the concrete? — this is obvious. These are the bricks from which the building is made. But why the intermediate, pictorial step? One answer is that it goes halfway — drawn objects are not tangible, they are more abstract. But there is yet another factor, no less important — pace. It is much faster to draw than to manipulate objects. It takes less time to draw 10 pegs on a page or on a personal whiteboard than to put 10 beads on a string. It is faster to draw 5 lines and cross out 3 than to collect 5 pencils and remove 3. And it is easier to share your work with the entire class — just raise your personal board with the picture you drew, so that everybody can see it.

Speed goes against my own grain — I am the meditative type. But having observed teachers over the years, I cannot but admit that the best teachers, not only math teachers, are fast. Speed energizes. Once, before I was interviewed on television, a member of the production team told me:

speak fast. The audience responds to your energy rather than to your message. "Arithmetic is done quickly," I keep telling the children. Inventing an arithmetic story? No princes or princesses (and if there are, they should act quickly). The story must be short: "I had 5 pencils, 3 broke, how many intact ones are left?" Fast and steady wins the race.

Thou Shalt Not Skip

DO NOT SKIP
SHOW, DON'T TELL
PRECISE TERMS
NAME YOUR TERMS
MEANING BEFORE CALCULATION
ARITHMETIC STORIES
STUDENTS INVENT QUESTIONS
PERSONAL BOARD
ESTIMATE
NOTHING IS TOO SIMPLE

"Begin with the concrete" is the first commandment. The second is — do not skip. Do not miss layers. Obvious? It was not for me when I first landed in elementary school. Before learning the right way from teachers,

I fell flat on my face many times. There is nothing more frustrating than standing in front of thirty children who do not understand what you want from them.

Here is an example — teaching the notion of "more by…" in Grade 1. Say, "John has 3 more pencils than Dina."

An obvious preceding rung in the ladder is plainly "larger," not "by." But there are more preliminary concepts, the first of which is "equal." Let us follow a lesson. The assumption is that the students sit in pairs, otherwise, pair them up before starting. The objects used are beads strung on skewers.

Teacher: Everyone, please put beads on your skewers so that you and your partner have the same number of beads — you will decide how many. Joe and Mary — please show us your skewers, how many does Joe have? Right, 5. And Mary? 5 again. By the way, let us see who is the laziest pair here — who put the least number? Lucy and Edith — you put 1 bead on each of your skewers. Nice. Can you get even lazier, and put less? Yes, putting zero beads on each skewer is okay too.

Now, back to Joe and Mary. Joe, I would like you to have 3 more beads than Mary. Can you do that? Nice, you added 3 beads. Now take off the extra 3 beads, back to 5-5. Mary — can you do something with your beads so that Joe will have 3 more? Yes, you can, just think. Indeed, you can remove 3 beads. Mary, come to the board and draw Joe's beads on the left side of the board, and yours on the right side. You can draw lines instead of beads.

And so on — getting slowly to the notion of "Having more beads," then "having 3 more beads," then "having 3 less beads."

Words: The Cement of Thought

We use words to hide our thoughts.

— Voltaire, a French philosopher, 1694–1778

Many a thing you feel you want to tell her. Many a thing she ought to understand.

— The nuns about Maria in *The Sound of Music*

Polonius: What do you read, my lord?
Hamlet: Words, words, words.

— Shakespeare's *Hamlet*

Not only humans have multi-level conceptual buildings. Our dog learned to open a gate to (1) enable her to reach a wall, (2) jump over it, (3) to roam the neighborhood, and (4) to find the cat food left by kindhearted neighbors. All these chain links dwell in her mind. So, what is man's advantage? The main answer is: words.

Words are not only for communication. They have at least three other important roles: focus, cementing conceptual structures, and memorizing.

Let me start with the first function. I do not believe in horoscopes (sorry), but I admit that they do play an important role: drawing people's attention to concepts. Here are some properties expected from me as a Scorpio, as they appear on some sites:

Scorpios are famously discreet and even secretive, with an air of mystery around them. Other personality traits include fearlessness, boldness (thanks to Mars), passion, creativity, and fierce loyalty.

Almost nothing fits me, but it makes me think — am I passionate? Loyal? These may be notions I have never considered before. Words draw attention.

The second function of words is cementing conceptual scaffoldings. Defining a layer in words makes it solid enough to support the next layer. Words are stakes in the rock, grips that ease further ascent. Principles that are put to words are ready for use.

Upon reaching elementary school, I believed the opposite — that words are not for children. Children need intuition. I couldn't be more off the mark. Children love words and are proud to use them. Some of my most enjoyable classes were on words and their origins. Why "multiplication"? Do you know a word close to it? Have you heard the word "multi-millionaire"? Why "numerator"? (It counts the number of parts). Why "denominator"? (It is the name of the part. "Denominate" is to give a name.)

Words also help remembering, they serve the "declarative" (as opposed to "episodic") memory. We often remember things because we were told about them, or uttered them ourselves. "Remember the time you put the ice cream in your shirt because dad plucked your shirt and told you there is enough room in there for more?" — of course I remember, after the 100 times you told me. Once events are put into words, they are easier to remember. If a fairy appears and offers me three wishes, one of them would be to be able to put to words the fuzzy insights I often have in my mind. My wife, a typical fisherman's wife,[1] may frown upon the waste of a wish, but I do not meddle with her wishes, let her not meddle with mine.

Once put into words, a thought can develop further. A new floor can be built on top of the verbal structure. Words enable moving on, they are sometimes an antidote to stagnation.

---

[1] In a Grimm brothers story, a fish offers a fisherman a wish as a reward for sparing his life. The fisherman's wife comes up with fancier and fancier wishes, and eventually she receives no wish at all.

Class Discussion

Considering his nationality and education, the Russian-Jewish psychologist Lev Wigodski should probably have ended his life in the Gulag.[1] Luckily for him, he died of tuberculosis in 1934 before the great Stalinist purges. It may be disputed whether he was indeed lucky, but his theories were undoubtedly fortunate — they won the support of the Russian establishment and consequently got recognition in the West as well. They fit the Soviet doctrine. "Everything is learned first socially, only later individually" was his dictum. The child learns by the mediation of adults and of the surrounding society. Indeed, common discussion is one of the most effective tools in class. It is easier to follow your classmates than the teacher — the feeling of "we are all together in this venture" pushes the whole class forward.

Unfortunately, in modern education, which promotes the individual, this is rarely used. The "Frontal lesson" (with the entire class) has become largely obsolete. Teachers are encouraged to teach individually, with each student learning at their own pace. Of course, this is extremely taxing on the teacher. Individual teaching can be great if you have the resources, but this is rarely the case. And it misses the power of class discussion.

---

[1]The vast system of labor camps in Siberia, that existed from the 1920s to the 1950s. Acronym of "Glavnoe Upravlenie Lagerei," meaning "Main Camp Administration."

Naming

What's in a name? That which we call a rose by any other name would smell as sweet.

— *Romeo and Juliet,* William Shakespeare

Indeed, the actual name itself is immaterial. But it is important to give a name, any name. It provides a handle on the concept. It enables not only communication but also thought.

Children are not afraid of names. In fact, they love them. Teaching addition, the components are called "addends." Hard? Not if the terms are taught by examples. In $3 + 4 = 7$, the numbers 3 and 4 are 'addends' and 7 is the 'sum'.

Some important terms:

In $7 - 4 = 3$, the number 7 is called "minuend," the 4 "subtrahend" and the 3 "difference" — frightening? Try it on your students and witness their pleasure.

In $6:2 = 3$, 6 is the dividend, 2 is the divisor, and 3 is the quotient.

The numbers 0, 1, 2, 3,... are "whole" numbers. We also call them "natural numbers" because they appear naturally, they are the first type of numbers we meet. The other types — fractions and negative numbers — were defined much later and were not entirely obvious.

The Personal Whiteboard

Class discussion is great. But how about practicing? Can it also be done with the entire class? Shouldn't every student do it in their own book?

In fact, it is possible to practice all together using a wonderful tool: the personal whiteboard. This is a very old invention. In the old days, school children held "slate boards," on which they wrote with chalk. Nowadays, there are whiteboards with markers. These can draw even the most sluggish of students out of passivity. Pose a common assignment to all students and then ask them all to show their work. For example, "draw $5 + 3$":

$$||||| + ||| = ||||||||$$

Encourage the students to replace the lines with something less boring, like flowers.

Or "please show us on your boards how you calculate 23 × 34." Yes, please — all stages of the calculation. Then ask the students, each in turn, to explain the steps of the calculation. This is more effective than a demonstration by the teacher. The children follow it less passively and with less stress.

Child Picasso's mathematical drawings

Laughter = Understanding

In Grade 1, after we join a group of 3 pencils with a group of 4 pencils, and conclude that 3 pencils plus 4 pencils are 7 pencils, I ask: how many are 3 apples plus 4 apples? 7 apples, they say. How do you know? We did it with pencils! Perhaps with apples it is different?

The ensuing laughter is a sign of understanding: the rules of arithmetic do not depend on the objects counted. $3 + 4 = 7$ is true for any type of object.

There are many ways to bring laughter about. One is asking trivial questions. Another is to repeat *ad nauseam*. What is 2:1? And 3:1? And 100:1? — they will laugh when it becomes too easy. Boring? Good. The students are happy to be on the victorious side. Laughter is a sign that some effort has become redundant. Victorious glee is there because once the enemy has been defeated, the energy we prepared for the war is no longer needed. What is $9 + 2$? $19 + 2$? $29 + 2$? $89 + 2$? $99 + 2$? $199 + 2$? A million and $99 + 2$? Laughter means the principle has sunk in.

Part 3

Kindergarten

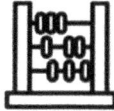

Why?

Genius is to see something everybody knows and ask a question nobody ever asked.

— Albert Saint Giorgi, 1893–1986, discoverer of Vitamin C, following Schopenhauer, a German philosopher, 1788–1860.

When I was four, walking with my parents in the paths of the kibbutz (communal settlement) in which we lived, I asked them "What makes the wind?" One of them, I don't remember who, improvised — "The tops of the trees move, and push the air." If you want to know why I became a mathematician and not a scientist (scientists deal directly with the physical world, mathematicians only indirectly), there is the answer.

In Grade 3, I almost re-discovered calculus, the mathematical branch whose starting point is that big changes result of the accumulation of small ones. "If all the air eventually comes out of the football, a bit of air must go out every second, shouldn't it?" I asked my mother. Her answer was "don't be stupid." This may explain why my research field is discrete mathematics, where the objects are separate, and there are no quantities that are "infinitely small."

Don't take these etiological explanations of my career too seriously. The message is how important it is to encourage curiosity. Some parents crave the label "gifted" for their children. Replace it with "curious," and I will agree. Curiosity is more important than knowledge, and pressure to be "gifted" stifles it. Encourage your child to ask "why." One way is to admit you do not know everything — let us look for answers together — for example, "what generates the wind?" — Let us look up Wikipedia.

This is as important in life as in science. Encourage independence. A small piece of advice: let your children win arguments. Of course, if they reason — not just "I want," but why. "I want more screen time, so as not to be behind with my friends." "I want to watch a film because I am too tired now to do anything else." Praise your child for a good argument. When I once refused a request of my then 10-year-old son Ziv and his friend, Ziv told him — "let me do the arguing, I am good at it." I was proud.

Here are a few "whys," some simple, some more complex.

- Why do we sometimes see the moon and sometimes not?
- Why are eggs oval, and why is one side "sharp" and the other "blunt"? (A ball-shaped egg would roll too easily. Dangerous for an egg.)
- Why does it rain in Israel only in winter?
- Why are the plants that bloom late in the season taller than those that bloom early?

- Why does the sun always rise on the same side?
- Why are there more right-handed people than left-handed? (An opportunity to learn about problems nobody has a convincing answer for.)
- Why is it possible to tame dogs and hard to tame cats? (*Hint*: in nature, dogs live in packs).
- Why are there so many right angles around us? (*Hint*: try to pack circles. Or, what happens to a heap of bricks that is laid slanted, not perpendicularly to the ground?)
- Why do we have two hands, two ears, and two eyes? (Careful: each has a different answer. For example, the two ears enable figuring the direction from which the sound comes by the difference in the time it reaches them.)

No less important is to ask "why" on customs and habits.

- Why do we eat on plates?
- Why is it advisable to walk on the left side of the road (or, in England, on the right side)?
- Why do we say "goodbye" upon departure? (So that the person knows we are still on good terms.)
- Why do you use blankets for sleep and not coats? (Encourage your child to try a coat to find out.)

Asking new questions about familiar things is the opposite of "déjà vu." Some call it "vuja de."

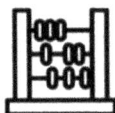

The First Operation: Naming a Unit

We live to be kind to others. Why are the others there — I have no idea.

— Thomas Dewar, a brewery owner

And his last words were 'thumbtacks, buttons'

— "The summary," Nathan Alterman, an Israeli poet,
about a sage who on his deathbed summarizes
what the world is composed of.

"Why are we here" is a tough question. An easier one is how we survive. The magic word is "order." Living creatures and plants accommodate to the order in the world. When a car approaches, we do not cross the road. Life is about order, and so is mathematics — it is the most efficient tool for discovering order. In particular, the arithmetic operations organize the world. 3 + 4 is about rearrangement: joining two groups, one of 3 pencils and another of 4 pencils, into one large group.

To arrange objects in some order, they should first be distinguished and named. This, and not addition, is the first operation. Partitioning the world to objects and naming them — "thumbtack," "button." The chosen object is called a "unit" and it is what we operate on: count, multiply, divide, and take a fraction of.

The unit can be a set of items, say 6 buttons. Repeating it 3 times is multiplication, 3 × 6, and dividing it between (say) 3 children is division, 6:3. In the decimal system, the basic unit is a group of 10. Ten of these form a new unit, "a hundred," ten of those form the unit called "a thousand," and so on.

Sorting by type is the first step in kindergarten. Look at the picture: what kinds of animals do you see? Fish, birds, lions, and giraffes. Have we forgotten anything? Yes, people! People are also animals. How do we know? They eat, they breathe, they move, like all other animals. Do flowers eat? Right — they drink water, but do they eat? Yes, but not through a mouth — through their roots.

Which animals have two legs? Right, birds and people. What do birds have instead of the two front legs? What do humans have?

A friend of mine uses bottles to teach sorting by type. He puts about 15 bottles on his desk: some big, some small, some filled with water, some empty, some half full, some with red caps, some with blue, and some with none. He then challenges the children to divide them into types. "There are 3 full bottles, 2 empty and 6 half-full," "5 with a cap, 10 without," and so on — you can run a whole lesson. The children will find criteria you haven't thought of, like "some are standing up, some lying on their side." Children are more observant than us grownups.

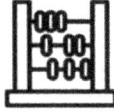

Directions

Pooh looked at his two paws. He knew that one of them was the right, and he knew that when you had decided which one of them was the right, then the other one was the left, but he never could remember how to begin.

— Winnie the Pooh, A. A. Milne

23 is not the same as 32. To tell them apart you must tell right from left. Directions are essential in arithmetic, in geometry even more so, and of course in everyday life.

Start as early as possible. Toddlers love learning directions through their body. When you give them a bath, tell them "I am rinsing your right hand, now your left hand, now right cheek, and now left cheek."

To solve the Pooh dilemma, apply the "divide and conquer" strategy. Start with just one side. Right foot, right hand, and right cheek. It is an opportunity to teach the body parts, also the less prominent ones: temple, thigh, ankle, eyebrows. Did I mention that children love names?

Ask what parts of the body come in pairs and which do not. To understand a concept (in this case: pairs), give also examples where it does not apply.

From age 4 and on, you can ask the child to give you directions: how do I get to the door? If there are obstacles, giving directions may be non-trivial.

At the age of 4 or 5, the child can understand the relativity of the directions. Ask two children standing one opposite the other — where is the window — to your right or to your left? Hmmm, one child claims right, the other left — who is right? Put two objects on the teacher's desk — say a bottle and a book and let one child stand on one side of the desk and another child on the opposite side. Ask them — what is on the right, the bottle or the book? In kindergarten, this may necessitate help: where is your right arm? And your left? Where is the book — on the side of which hand? Why are the students' answers different? This is a good way of overcoming egocentricity. Different people see things differently.

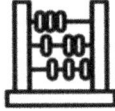

Before and After, Smaller and Bigger

"To the left of… to the right of…" naturally leads to other relations. Here is a short list.

1. Temporal relations: before and after. Newer and older. In "Here we go round the mulberry bush," which word comes first, "we" or "mulberry"? What is the first word of the song? What is the last? I am performing a few actions — I raise my right hand, raise my left hand, protrude my tongue — can you tell me what I did first? And second? And third? What did I do last?
2. Can you repeat what I did, from last to first?

    Can you tell what you did today, from the morning until now? Can you tell it in reverse order?

    What day comes before Saturday? And before? (For the clever) — can you tell the day in one week's time? And in 8 days? Pair the children and in each pair, ask one child to perform a few actions, and the other to describe what the first one did. Then, describe the actions in reverse order.

What is earlier, the chair or the table? Right — tables and chairs are not "before" and "after." So, what things *do* we say are "earlier" and "later"? The children may answer: "about actions." But we also asked "earlier" and "later" about Saturday and Friday — are the days things that we do? No, we don't do Saturday. It is not an action. It is something that happens. We say "later" and "earlier" about things that happen. Namely, about events.

What is the first thing you remember? What was your first pet? And your first friend?

This can lead to conversations on time. How do we measure time? Can you tell me something that takes long? Yes, our lessons. Longer? Yes, the whole day. Longer? What is longer than an hour? A day, a week, a month, a year. And shorter? A minute, a second. Shorter than a second? Clever students will answer — half a second.

3. Bigger-smaller. There is no end to the activities you can do around these concepts. A warning I learned from teachers: do not use the height (certainly not weight) of the children as examples. This is often a sensitive issue.

It is a good opportunity to teach "big, bigger, biggest."

What is bigger, the moon or the sun? They look more or less the same size, don't they? But the sun is much farther, 389 times farther — what do you think, who is bigger? Here is a picture of the tall buildings in New York — the picture is very small, how come the buildings are big?

Here are some more relations

4. Longer–shorter, wider–narrower.
5. Lighter–heavier
6. Older vs. newer. Tanya — who is older, you or your brother? You are older. So, is he newer? No, he is younger. With people (and animals) we say "young," not "new."
7. Tastier — less tasty (there are non-absolute parameters — who prefers chocolate to ice cream? And who prefers ice cream?).

Equations in Kindergarten

I put 5 pebbles on the ground, and ask four-year-olds to count them. Now shut your eyes "tight," I tell them. You should see how tight their eyelids are. I then cover 2 pebbles with my hand, ask them to open their eyes, and ask — how many pebbles do I have under my hand? Let us check. Correct! 2. How did you know?

The kids solved an equation: $3 + x = 5$. Next, I cover all 5 pebbles — how many did I cover? They laugh — it is like asking "What was the color of Napoleon's white horse?" Now I do not cover any pebbles — again they laugh and learn the notion of "zero."

Then let the children pose themselves such riddles. One child hides some pebbles and asks his or her friends to shut their eyes "tight."

Yet another way to form an equation: build two towers of identical cubes, one with 6 blocks, the other with 3. How many blocks do I need to add to the lower one so as to make them of equal height?

I was surprised to find how difficult the following question was for three-year-olds. I put two building blocks one on top of the other, the bottom one short, and the top one long. Then I put a long building block on the ground and asked what I should add to make the two towers equal. This is "A + B = B + A" — the commutative law. It may take them some time to find the answer.

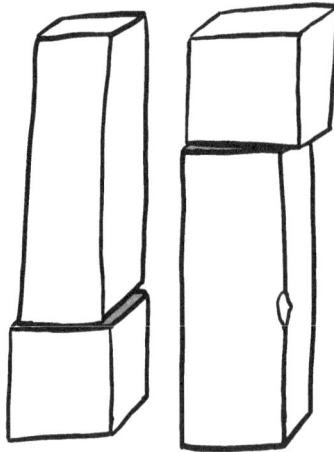

This is a good opportunity to learn conditioning: "if." Here are 5 pebbles on the table. What will happen if I take them all, how many will be left? (always start with the simplest). What will happen if I take one? How many should I take to leave none?

Part 4

More Teaching Principles

Reversal*

The great violinist of the end of the 19th century, Joseph Joachim, met the child prodigy Arthur Rubinstein, later to become one of the greatest pianists of all time. Joachim played the theme of Strauss' Blue Danube and asked the child to play it backwards. Little Arthur succeeded, and Joachim took him under his patronage.

Geniuses like this are rare. Even repeating the words of a short, familiar poem backward is next to impossible. Try it with "Mary had a little lamb." It is non-trivial but rewarding. Once you succeed, you will know the words much better. In Grade 1, count back from 10 to 1, then from 20 to 1. Then continue to zero — an opportunity to get acquainted with this number. Try asking if it is possible to continue, probably some students will know what "minus 1" is. What follows (−1) in the sequence 3, 2, 1, 0, (−1)? Some will know: (−2), (−3),…

Why is this so important? It is a way of achieving "deautomatization" — a term coined by the Russian literary theorist Viktor Shklovsky (1893–1984), who was, by the way, the son of a mathematician. Forward, you count automatically. Backward, you must think.

Next, reverse actions.

---

* Reminder: 🖾 icon stands for "a thought or a teaching principle."

Teacher: Please stand up — what is the opposite? Go one step forward — what is the opposite? Jenny — please go three steps forward. How will you get back to where you started? What is the opposite of adding 3? And of subtracting 3?

In Grade 2, you can ask: Jean walked three steps forward and then four steps to the right, how should she return? Draw it on the board.

Do you know any other actions that have the "opposite"? What is the opposite of dressing? Of putting on shoes?

Next, opposite properties.

What is the opposite of "beautiful"? And of "ugly"? Of "tall", of "low"? Of "far," of "near"? Can you come up with more examples?

Next, ask what the opposite of "table" is — no, it is not "chair". "Table" does not have an opposite. This is an opportunity to discuss the difference between "property" (do not shy away also from "adjective") and "noun." It may be a good opportunity to teach these terms. Of course — slowly and with lots of examples. Adjectives have opposites, nouns don't.

What is the opposite of "big"? And the opposite of the opposite of "big"? And three times "opposite"? And 10 times? And 11? And 100 times?

My First Equations: Using Reversal

I ask a student standing on a staircase to climb up 3 stairs — how will she return to the place she was before? Of course, go down 3 stairs. Now I ask: a girl climbed up 3 stairs and reached stair number 100. Which stair was she standing on when she started? We know — going back to where she started is done by descending 3 stairs, $100 - 3 = 97$.

We are solving an equation: "a number $+ 3$ is 100, what is the number?" The main tool in solving equations is reversal. A number times 3 is 39, what is the number? The inverse of multiplication by 3 is division by 3. Answer – $39:3$, which is 13.

The four operations are arranged in two pairs of inverse operations, addition–subtraction, multiplication–division.

Things become more interesting when there are two operations. A number is multiplied by 3, and then 5 is added to it, resulting in 32. What is the number?

I ask the children: in the morning, in what order do you put on your shoes and your socks? Indeed, socks first, shoes second. Going to bed, in what order do you take them off? Yes, in reverse order. Last worn, first taken off. So, how do you reverse the two operations of multiplying by 3 and adding 5? To go back to the original number, you first subtract 5 from 32 (to get 27) and then divide by 3 to get 27:3 = 9.

A special case is "the whole by the fraction":

$\frac{2}{5}$ of a number is 100. What is the number (the "whole")?

Solution: Taking $\frac{2}{5}$ of a number means dividing it by 5 to get 1/5 and then multiplying the result by 2. To get back to the original number, you divide by 2 (to get one-fifth) and then multiply by 5. So, the number is

$$100: 2 \times 5 = 50 \times 5 = 250.$$

Role Reversal

In the usual ping-pong game, the teacher asks and the students answer. The students are passive, they only react. It is important to occasionally reverse roles. The students should invent questions, concocting exercises, or arithmetic stories. Ask the students: can you find an addition exercise whose outcome is 100? Is there more than one possibility?

Encourage them to find the simplest exercise. Rather than solving equations, ask them to compose equations.

Can you come up with an equation whose solution is 3? Yes, $x + 5 = 8$. What is the simplest such equation? Perhaps, $x + 0 = 3$. Or perhaps even $x = 3$ — is this really an equation? Yes, though it is its own solution, remember Napoleon's white horse?

Find stories in which the required operation is 3×4. Compose a story requiring a million divided by 2.

The person at the steering wheel gets to know the road best.

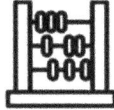

Conservation and the Commutativity
of Addition

In Grade 1, I ask a student to demonstrate with "flowers" (popsicle sticks) $3 + 4$. She holds 3 sticks in one hand, 4 in the other, and joins them. Then I ask her, before joining the two sets, to cross her hands — right becomes left and left becomes right. Has the number changed? Of course not. So,

$$4 + 3 = 3 + 4.$$

This is the same cat, although in different positions.

Rearranging the pebbles does not change their number.

This is called the "commutative law" of addition. It is a **conservation law**. Such a law says that while one thing changes, another stays put. A cat, moving from the chair to the bed, still remains a cat — this is the preservation of identity. If we move around pebbles on the desk, their number doesn't change. This is the conservation of number. If we cross our hands holding the "flowers," their number remains the same.

There Is No Such Thing as "Too Simple"

The Hungarian mathematician George Polya (pronounced "Poya") wrote a book that quickly became the bible of principles of mathematical thinking: *How to Solve it*. It is a great book, but do not expect to become certified problem solvers by just reading it. Reading a book on swimming will not make you an Olympic champion. But thought principles are also teaching principles, so you may become better teachers.

George Polya (1887–1985).

One of the principles in *How to Solve It* is — if a problem is hard, look for an easier version. Always strive for the simplest case. There is no such thing as "too simple." My young daughter never read Polya's book, but she came up with the same principle. Whenever I asked her something she didn't know, she would say, "ask me a simpler question." She was not dodging the question, but rather asking for an intermediate step.

Part 5

Operations in Search of Meaning

Any life is tolerable if it has meaning.

— Viktor Frankl, 1905–1997, *Man's Search for Meaning*

Meaning before Calculation

There is a common expression in the teaching of mathematics: "teaching artifacts," or "means of demonstration." We first calculate $4 + 3$ and then show it with pencils and pebbles. This reflects poor understanding. Three flowers joined with 4 flowers are not an illustration for "$3 + 4 = 7$" — they are the thing itself. Just as a cat is not an illustration for the concept of "cat." The equality $3 + 4 = 7$ is an abstraction from 3 flowers joined with 4 flowers. Meaning should come before calculation. The only way to calculate $3 + 4$ is to play with concrete objects.

Does this sound elementary? For many years, it was taught in the wrong order, in many countries, including my own. Thanks to the struggle of some persistent mathematicians, this practice has changed.

Begin with objects. You can call it "arithmetic shows." The next step is arithmetic stories. Tell a story about $3 + 4$, about 3 times 4 ("How many legs do 3 chairs have?"), about $4-3$. Students who are acquainted with arithmetic stories from Grade 1, and even more importantly, know how to make them up, will find the dreaded "word problems" of higher grades less intimidating.

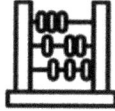

The Curriculum in Arithmetic

What should students at elementary school know in arithmetic (namely, about numbers)? The answer is astoundingly simple: the five operations (remember? To the four staple operations, we added a fifth, that of forming a unit).

Just five operations! Why does it take six years to learn them? The answer is — because they have two facets: meaning and calculation. And neither of these is simple. Meanings are often hidden, and calculation — well, you probably remember how you struggled with it as a child.

The operations, we said, are re-organization of the world. Addition is joining two sets, subtraction is taking away one set from another. But then — why numbers? Putting a piece of cheese on a slice of bread reorganizes them — isn't this an operation? Why is $9 + 3 = 12$ more interesting than that? Because we do not only join two sets, of 9 and of 3 objects. We then rearrange them as $10 + 2$.

This has to do with the way we represent numbers — the decimal system. Calculating an exercise is not just finding the result. It is finding the **decimal form** of the result.

So, in elementary arithmetic you study two topics:

1. The meaning of the operations, and
2. The decimal system.

We start with the first.

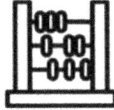

Addition and Subtraction: Dynamic and Static

Addition has at least three meanings, exemplified in:

1. There were 4 birds on a branch, 3 joined them for a night's rest — how many are there now?
2. In a vase, there are 4 yellow flowers and 3 purple flowers. How many are there together?
3. Joe has 4 markers. Jane has 3 more than Joe. How many does she have?

The first type is called "dynamic," something changes, the added objects were not there before. The second is called "static," and the third is "comparison." Should we bother the children with these names? Yes, by all means. They will take pride in mastering the word "dynamic," and their vocabulary will be enriched. Grade 1 and Grade 2 students should be able to recognize the type, and they should be able to invent stories of each type on their own.

The same goes for the reverse operation, subtraction. Here too there are three types, exemplified in:

1. There were 7 birds on a branch, 3 flew away, how many remain?
2. There are 7 flowers in the vase, purple and yellow. There are 3 purple, how many are yellow?
3. Jane has 7 markers, Joe has 3 fewer, how many does he have?

The three types are not completely different — for example, in type 2 one can say "in your imagination, remove the 3 purple flowers. How many remain?"

Multiplication = Counting

Students in English-speaking countries have an advantage when studying multiplication — English uses the right term for it, "times." It conveys the meaning. Multiplication is repetition. 3×4, in words "3 times 4," means taking a group of 4 apples and repeating it 3 times.

But repeating is just counting! $3 \times$ apple is repeating the apple 3 times, so it is just 3 apples! In a formula, $3 \times$ apple = 3 apples. The multiplication sign is redundant. Indeed, in algebra $3 \times a$ is denoted by $3a$. It economizes in notation, but more importantly, it conveys the meaning.

Counting is a special case of multiplication.

The flowers have 3×4 leaves

Distribution: Open the Box before Multiplying, or After?

What is 3×24? Everybody knows — we have 3 times 2 tens, which is 6 tens, plus 3 times 4, together 72. Without noticing, we used *distribution*. $24 = 20 + 4$, so the product is $3 \times (20 + 4)$. Distributivity tells us:

$$3 \times (20 + 4) = 3 \times 20 + 3 \times 4.$$

Proof:

$$3 \times (20 + 4) = (20 + 4) + (20 + 4) + (20 + 4)$$

Which, by commutativity of addition, is

$$(20 + 20 + 20) + (4 + 4 + 4) = 3 \times 20 + 3 \times 4$$

The parentheses are like a box — we multiply its entire content by 3. This multiplies both the 20 and the 4. If we open the box after the multiplication, we have 3×24. If we open before, like an impatient child with his gifts, and then multiply the terms separately, we have $3 \times 20 + 3 \times 4$.

It is an opportunity to formulate a law in algebraic terms:

$$3 \times (a + b) = 3 \times a + 3 \times b$$

What are a and b? Grade 5 (and on) students will know: any numbers. Or even any objects — I like substituting apples and bananas, suiting the a and the b. Using letters to denote general numbers is called "algebra." Tell the students they learned algebra — they will be proud.

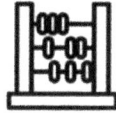

$4 \times 3 = 3 \times 4$ Is Not an Axiom

I told you about the way my daughter Inbal proved that $2 \times 3 = 3 \times 2$, using her fingers. You probably do not have enough fingers to show in the same way that $4 \times 3 = 3 \times 4$, but you can draw instead. Look at the following circles arranged in a rectangle.

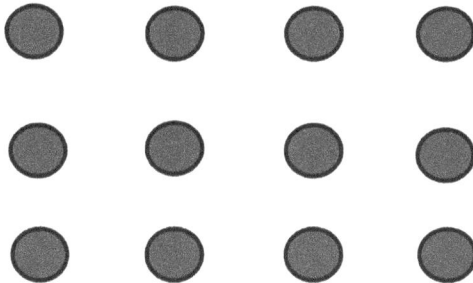

There are two ways of counting the circles: there are 3 rows, each containing 4 circles, so it is 3×4. Another way: there are 4 columns, each containing 3 circles, so it is 4×3. If you view the circles as fingertips, this was precisely Inbal's proof.

The commutativity of multiplication is not an axiom, it is a provable theorem. Here is an alien with four hands proving it.

An alien showing that 3 times 4 is 4 times 3

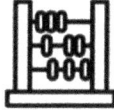

Two Types of Division

I am not sure how new all the above was to you. But division, our next topic, has something new to offer: it has two substantially different meanings.

Ask a student for a story about 6:2, and you will probably get something like "A mother divided 6 cookies among her 2 children, how many cookies did each get?" This is called "sharing division." You divide into equal parts and ask how large each part is.

But there is yet another story:

Mom divided 6 cookies among her children, and each got 2.
Here I stop — what haven't I told you? The students know —
How many children does she have?

The first question is — if there are two parts, how large is each? The second question is — each part is of size 2, how many parts are there? In other words: how many 2s go into 6? This gives this category its name: "containment division." Best is to use also more suggestive names: "how large is each part" for sharing division, and "how many parts" for the containment division.

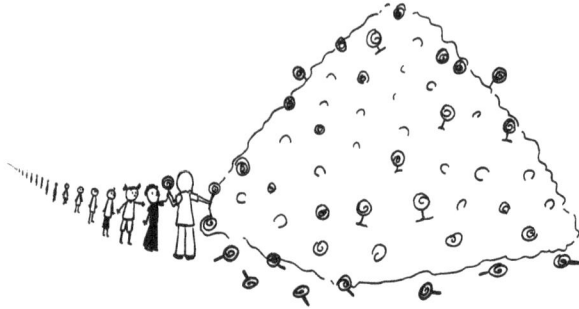

A million candies are divided, and each person gets 1, how many people got candies?

The sharing version corresponds to $2 \times 3 = 6$, saying that 2 sets of 3 are 6.

The "how many parts" is $3 \times 2 = 6$, 3 sets of size 2 give 6.

So, the two types are all about the commutativity of multiplication.

Sharing division is more familiar, but the other meaning is no less important. In particular, for calculation. To calculate 91:7, you use the meaning of "how many times does 7 go into 91?"

In distributing cards, both meanings can be seen. Distributing 91 cards among 7 players, you have a first round of distributing 7, then a second round, and so on — when you finish you find that there were 13 rounds — 13 times 7 went into 91. "How many rounds" is "how many times does 7 go into 91." From the other point of view, "how many cards does each participant get" is "how large is each part."

Another place where the "how many parts" is essential is division by fractions. What story can you tell, to show that $10 : \frac{1}{2} = 20$? A "how many in each part" story is — "there are 10 cookies, divided evenly between half a child. How many does each child get?" It does make sense — if half a child gets 10, then each child (namely, one child) gets double — 20. But it is much more natural to ask "how many halves go into 10?" — each cookie contains two halves, so 10 cookies contain $10 \times 2 = 20$ halves.

A million cookies are divided. Each person gets $\frac{1}{2}$ a cookie. How many people are there?

Of course, we start with "how large is each part," because this is the type encountered in everyday life. Children are used to dividing things fairly. Start with dividing 4 popsicle sticks between 1 child — combining two teaching rules: "start with the trivial" and "make them laugh." Then divide 0 sticks among 4 children — just as funny. Let a child with 0 flowers divide them theatrically between two children (they should say "thank you"). Then 4 objects among 4 children — almost as funny. If the students are seated in pairs, let every pair choose what to divide between them — some will choose 10 objects, some zero objects. Ask one pair to divide between them their 20 fingers — laughter guaranteed.

Then tell them it is possible to divide also single objects — a cake, or a blank page.

Ask a pair of students to divide 5 "flowers" between them. What do you do with the remaining 1 that was not distributed? Let them discuss this.

Division is a rich subject. I fondly remember many enjoyable classes on it.

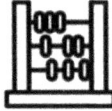

The Ultimate Crime: Dividing by Zero

The Catholic Church counts seven sins penalized by death — "mortal sins":

pride, greed, lust, envy, gluttony, wrath, sloth.

If death penalty for these sins was indeed delivered, not many would be spared, perhaps only a few Buddhist monks. The list was not concocted by God, but by his representative on earth, Grigorius the Great, who was pope between 590 and 614. Math teachers add an eighth sin: dividing by zero. Ten divided by zero (10:0) is unthinkable blasphemy. Students are in such awe, that they are wary also about dividing zero by something, though 0:10 is perfectly legal — each of the 10 participants gets 0 (it doesn't matter what).

In fact, division by zero is not important. It should be mentioned briefly, then skipped. I usually devote one lesson to it and go on to bigger and better things. In this lesson, we do two activities, that clarify the meaninglessness.

The first activity is about the "how large is each part" sense (sense-lessness, rather). I call a student (call her Joyce) to the front of the class, give her (say) 6 sticks, and ask her to divide them between 6 students, then between 3, between 2, and between 1. Then I summon 0 students to the front of the class (not forgetting to greet them for joining us), and I ask

Joyce to divide the sticks among "them." She is perplexed — there is nobody to give the sticks to. Mission impossible.

Then we do it with the "how many parts" division. I ask Joyce to distribute her 6 sticks so that everybody gets one stick — how many will get sticks? Of course, 6. Then 2 — she calls students and gives each 2. There are 3 students, 6:2 = 3. Then I ask her to give 0 to each student — how many will there be? She walks from student to student in the class, and with a theatrical gesture gives 0 to each. The receivers thank her heartily. After a while, I ask — when will the "flowers" be exhausted? When will she finish the division? Of course, never. 6:0 is not doable.

The theatrical gesture and the "thank you" for the zero sticks each student receives will not be quickly forgotten and with it the senselessness of 6:0.

Nobody wants them. And I am supposed to go home only when they are finished

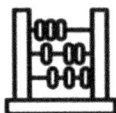

Remainder: When You Cannot Split

Give two students 7 cookies and ask them to split them evenly. Each gets 3 and one cooky is left. What should we do with it? Offer them a knife and they will split it in two — each gets $3\frac{1}{2}$.

Now do the same with 7 markers. The one remaining marker is undivisible. What do you do? There is no choice, it remains on its own. So, we call it "remainder." The remainder is the non-divided part.

We write it 7:2 = 3(1). The remainder is written in parentheses.

Another way of writing it is $7 = 2 \times 3 + 1$. Each child gets 3 and 1 is left over.

One case of the remainder is well-known — in division by 10. What is 23:10? 20:10 is 2, and 3 remains non-divided — it is the remainder. Similarly, 12345:10 is 1234(5), because $12345 = 1234 \times 10 + 5$. The rule: the remainder upon division by 10 is the rightmost digit.

What is the remainder of 12345 upon division by 100?

(answer: 45, can you explain why?)

I once wanted to teach the notion of "remainder," in Grade 2. I gave 4 students 6 coins of 10 cents and asked them to divide the coins evenly between themselves. They outwitted me: one of them took out 5 cents coins, exchanged 4 of them for two 10 cents coins, and then everybody got 15 cents.

Parity

Suppose the light in your room is off. What happens if you press the switch? And if you press again? And again? What happens if you press 100 times?

Let us write down the numbers that after pressing them the light is on: 1, 3, 5, 7,…. And the numbers that if you press the switch that number of times, the light is off? 2, 4, 6, 8,… have I forgotten an important number? Yes, 0. What does it mean to press zero times? What happens if you press it a million times (don't try it at home)? — the light is turned off because a million is an even number. How do you know it is even? "The last digit is 0, which is even," the students say. Is there another way? Yes, you can divide it into two equal parts, a million is 2 times half a million. But how do you know that half a million is not a fraction, but a whole number ("integer")? It is 500 thousands — a half of a thousand thousands.

> Harry — were you born in this classroom? No? Good. I didn't think you were. Have you ever entered through the window? No? Good — Harry has only ever entered through the door. Kids — is it possible that Harry went precisely 100 times through the door?

If the students are confused, ask them — is it possible that Harry went through this door twice?

(If a question is hard, ask it on small numbers.)

No — after one time he would be in, so after two times he would be outside. How about 3 times? Yes, we know he went through this door many times, not 3. But suppose you don't know that — could he? Yes, he could. After 3 times he is in.

After some discussion, you will reach the conclusion that Harry went through the door an odd number of times. Is 100 an odd number? No, it is even, so he couldn't have gone through the door 100 times.

"Yesterday I didn't not go to school," I tell the children. "Where I was?" After some thought they know — I did go to school. "And last Thursday I didn't not not go to school — did I go or not?" Of course, I didn't go to school. If it is "not not not not… not" 10 times? Well, after one not, it is "no," after 3 it is "yes," and after 3, it is "no," 4 "yes" — the children will get the pattern. 10 is an even number, so 10 consecutive "nos" is a "yes."

Aren't these stories similar? We call two things that behave the same and are only different by name "isomorphic." Let us dwell a bit on this notion.

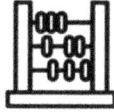

Isomorphism

Here is a two-person game. Each chooses in turn a number between 1 and 9, which hasn't been chosen so far. The winner is the player who has three numbers that add up to 15.

Here is an example of a game between players A and B:

A-5, B-3, A-2

Now A is threatening to choose 8, which would complete the pair 5, 2 to 15, so to prevent it, B has no choice: B-8.

Now B threatens to add 4, to complete the pair 8, 3 at his disposal to 15, so A has no choice: A-4.

Now A has two threats — to complete 4 and 9 by 2, and to complete 4 and 5 by 6. B can't prevent both, so A wins.

After a bit of experimentation, some students may notice: we are actually playing tic-tac-toe, on the magic square. The rows, columns and the two diagonals contain all possible triples summing up to 15, so the players are actually trying to complete a row, column or diagonal, just as in tic-tac-toe! The two games are the same — they are isomorphic.

Two games that look different are actually the same. They are "isomorphic."

The rows, columns, and two diagonals of the square represent all triplets of numbers between 1 and 9 that sum up to 15. Instead of choosing numbers, you can choose squares to put in them X or O.

Beauty, we said, is sensed when hidden order emerges. This is an example — on first encounter, the two games look different. The smiles on children's faces when they understand the connection tell us they sense beauty.

But let us return to parity.

Here is the door story in numbers. What is $1/\left(\frac{1}{4}\right)$? Indeed, 4. What is $1/\left(\frac{1}{1/4}\right)$? It is 1/4. What do you get if you continue doing this operation, of taking 1 divided by the previous number, 100 times? After an even number of steps, you get 4, after an odd number of times 1/4. The same logic — the two stories, this and passing through a doorway, are in essence the same. They are "isomorphic."

Average

Joe has 3 dollars and Jane has 5. How much do they have on average? Of course, the middle between them, 4. There is another way of saying it: if they join forces, they have $3 + 5 = 8$ dollars, and if they now share evenly, each of them gets $8:2 = 4$.

The second formulation can also be applied to three numbers: Alice has 10 dollars, Bob 2 and Carol 12. What is the average? Altogether they have 24 dollars, and when divided equally they will get each $24:3 = 8$.

Ask the students:

> We are 30 students in the class. If each has 10 cards, how many do we have on average?

Of course, 10 (this is the "juggling one ball" stage — a principle we shall later pursue).

Now let Jane give John 2 cards, how many do we have on average now?

True — still 10, because the total number of cards remained the same.

Here is a nice application of the average. What is the sum $1 + 2 + 3 + 4 + 5 + 6 + 7 + 8 + 9$? The average of these numbers is the middle between 1 and 9, which is $\frac{1+9}{2} = 5$. We have 9 numbers, that are on average 5, so their sum is $9 \times 5 = 45$.

Exercise: Use the same logic to find $1 + 2 + 3 + \cdots + 97 + 98 + 100$.

Answer: $100 \times 50\frac{1}{2} = 5050$.

This is the subject of the most often-told story in mathematics. The class of seven-year-old Carl Friedrich Gauss was asked to calculate this sum. The teacher was hoping to keep the students busy for an hour, so he could have some rest. Little Gauss promptly came up with the answer. He got it another way: he split the sum into pairs: $1 + 100, 2 + 99, 3 + 98....$ There are 50 such pairs, and each sums up to 101, so the sum is $50 \times 101 = 5050$.

Part 6

There Shalt Be Order:
The Decimal System

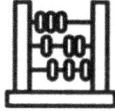

Collecting Tens

What are the most important mathematical discoveries of all time? Four are undisputable:

1. The notion of the number.
2. The decimal system, which organizes the world in tens.
3. The "place value system" for writing numbers, in which each digit counts the number of appearances of some power of 10. In 234, for example, the 4 counts ones, the 3 tens, and the 2 counts hundreds.
4. Calculus: the mathematics of continuous change. It was invented by the Greeks and developed in the 17th century.

We discussed the first achievement in the chapter "The First Operation," and I will later say something about the fourth. Here I want to dwell on the second and third. One is an ingenious way of organizing large sets, and the other is an ingenious way of representing numbers.

First, a small appetizer, intended for grownups, not for children. Question: Is the number 123456789 divisible by 3?

In grade 3 or 4, you learn the rule: sum up the digits, and check if the sum is divisible by 3. We shall later see why this criterion works. The sum is $1 + 2 + 3 + 4 + 5 + 6 + 7 + 8 + 9$, which we found in the previous chapter to be 45. This is divisible by 3, since 45:3 = 15 (another way: the sum of the digits is 4 + 5 = 9, which is divisible by 3). The rule is:

If the sum of the digits is divisible by 3, then so is the number itself.

So, the original number is divisible by 3.

Something seemingly impossible happened here: a free lunch. To check divisibility by 3 directly, we should take 123456789 objects (say, matchsticks) and check if they could be divided into groups of 3, with no remainder — a tremendous amount of work. How did we manage to save it?

Indeed, there is no free lunch. Some work has been invested here. In fact, a lot of it. It was put into the organization of the matches in groups of tens, hundreds, thousands, and so forth. In this number there are 1 hundred millions, 2 tens of millions, 3 millions, 4 hundred thousands, … and eventually 9 un-gathered matches, namely 9 ones. It would take a lot of work to do all this gathering, work that is hidden behind the representation of the number.

This is akin to the work invested in the preparation of dictionaries and phone books. A lot of work is put into ordering the words by alphabetical order (before the invention of the computer there were people who actually did it manually). This is rewarded many times over, in the ease of finding specific words. The decimal system is similar: you invest a lot of energy in arranging the objects in tens, hundreds, and so on, but once you have done it the world is yours — you can handle the number, find its properties (like divisibility by 3) and perform arithmetic operations.

Being accustomed to all this, it is easy to forget how clever the decimal system is. Without the concise representation of large numbers it avails, modern technology would not be.

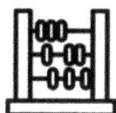

A Lesson on the Decimal System

Let's start at the very beginning, a very good place to start.

The Sound of Music, a musical

Let's start at the very beginning. That is, concretely, collecting tens of objects. Each student collects groups of ten popsicle sticks, and shows the class — I have 3 groups of 10 and 7 un-gathered sticks. We write it as 37. The next stage is drawing objects and circling tens of them on the personal boards. See to it that at least one student has less than 10 objects on his or her board — there is no gathering to do in this case.

After this activity, we fill the class whiteboard with small "pegs" — hundreds of them. The best is to ask students to come to the board to help. I ask for a guess — how many pegs are there? We write the guesses on another board, and later check who came closest. Then students come to the board — as many as can fit in its width. They stand on chairs, and circle tens of pegs. They should avoid, as much as possible, leaving "islands" of un-circled pegs. There are lots of tens, so after all pegs have been circled, apart from a number smaller than 10, I summon more children and ask them to circle tens of tens. These are hundreds. We count the hundreds, say 7 of them, then count the loose (un-circled) tens, say 3 of them, then the un-circled pegs — say 5, the number is 735. We say the number aloud, pointing at the 7 hundreds and 3 tens.

In Grade 3, we perform a feat — counting thousands of objects. The secret is cooperation. I distribute a large number of matches (or toothpicks) among the children, say 100–200 each, and ask them to collect their matches to tens and tie each 10 in a rubber band. Then every row collects the loose matches to tens, and once this is done, they collect tens to hundreds. We write on the board the numbers of each row. Finally, we collect all heaps together and gather tens of hundreds to thousands.

All this takes 20–30 minutes, after which the students have a feeling for what a thousand is or even several thousands. They will have had a hands-on experience with the decimal system, and they will never forget the time they counted 5000 matches.

The Second Most Useful Mathematical Invention (After the Number)

Archimedes, the Sicilian genius, made a discovery that could have changed the course of science. It did not, because he kept it to himself. He was probably unaware of its importance, and it had to wait some 800 years to be re-discovered. Gauss, two millennia later, lamented the loss — "who knows where science would be today had Archimedes spread the word." Gauss, by the way, knew a thing or two about keeping one's discoveries to oneself. In his later years, he rarely published. Eric Bell, a historian of mathematics, estimated that had Gauss published, mathematics would have advanced by some 50 years.

What was this exceptionally significant discovery? It is something every child takes for granted nowadays — the place value system for writing numbers. The place of the digit in the number determines its value. In 314, the 4 denotes units, the 1 is a ten, and the 3 counts hundreds. Each step to the left multiplies the quantity the digit signifies by 10.

The idea of gathering tens and then hundreds and so on is old. The ancient Egyptians passed it on to the Greeks, who gave it to the Romans. The Chinese used it, too. The Babylonians gathered in groups of 60, which is of course harder to apply — can you tell at a glance the difference between 58 and 59 pebbles? So, they divided each "sixty" into fives and tens.

The big hurdle was writing the numbers. The Roman way is as cumbersome as can be — XCI denotes 91 because C is 100, X is 10 and is to the left of the C, so it should be subtracted. You need infinitely many letters, for larger and larger numbers. In the place-value system, you need only 10 digits, since each can play infinitely many roles, according to its place in the number.

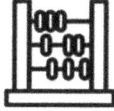

Why Didn't the Romans Have a "Zero" Symbol?

In Roman numerals, the number 305 is written CCCV. There is no need to put a zero, to signify that there are no tens. The absence of X's (tens) speaks for itself. In 305, the zero is needed, since in 35 the 3 signifies tens. The zero is a "place holder," or rather — place pointer, declaring "this is a place in the number, that should have denoted the tens."

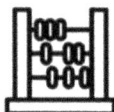

Divisibility by 3

Be wise, generalize.

Remember the rule for testing divisibility by 3?

(R3) If the sum of the digits of a number is divisible by 3, then so is the number itself.

The number 111111 (having 6 ones) is divisible by 3, since the sum of its digits is 6, which is divisible by 3.

Why is this rule true? Even those who learn it in school often do not know the reason. Strangely, sometimes it is easier to prove something more general, because it hits the nail on its head. It is the "right" fact. In this case, the "right" rule is:

(R3GEN) A number and the sum of its digits have the same remainder, upon division by 3.

Rule R3 is the case that the remainder is 0, namely the number is divisible by 3.

Example: in 100 the sum of the digits is 1, so the remainder of the sum, when divided by 3, is just 1. Indeed, the remainder of 100 upon division by 3 is 1, because $100 = 3 \times 33 + 1$. So, the rule is true in this case.

Why is R3GEN true? It follows from the fact that 9 is divisible by 3, and so are 99, 999, etc. Let us look at an example — as you will see there is nothing special in it. It clarifies the general idea.

Consider the number 5782 (I did not choose it arbitrarily — it is the number of the Jewish year, at the time of the writing of this book).

$$5782 = 5000 + 700 + 80 + 2 = 5 \times (999 + 1)$$
$$+ 7 \times (99 + 1) + 8 \times (9 + 1) + 2.$$

Using distribution (by which $5 \times (999 + 1) = 5 \times 999 + 5 \times 1$), this is $5 \times 999 + 7 \times 99 + 8 \times 9 + 5 + 7 + 8 + 2$.

But 999 is divisible by 3, and so are 99 and 9, so $5 \times 999 + 7 \times 99 + 8 \times 9$ is divisible by 3. So, the remainder of 5782 (which as noted above is the remainder of $5 \times 999 + 7 \times 99 + 8 \times 9 + 5 + 7 + 8 + 2$), is just the remainder of the rest, namely the remainder of $5 + 7 + 8 + 2$ upon division by 3. Namely, the remainder of the sum of the digits. This is rule R3GEN.

Since 9, 99, 999, etc. are also (obviously) divisible by 9, the same rule, with the same explanation, goes for divisibility by 9. The number 9234 is divisible by 9, since $9 + 2 + 3 + 4 = 18$, which is divisible by 9. The remainder of the number 555 upon dividing by 9 is the same as that of $5 + 5 + 5 = 15$, namely 6 (since $15 = 9 + 6$).

Part 7

The Checkered Notebook: Calculation

The Calculator Debate

Why learn how to calculate? We don't often meet in life something as complicated as 234×56, and if we do, we can use a calculator. The chances of being stranded in the desert without a calculator, while being required to calculate 234×56, are slim.

This logic prompted educationalists of the 1990s to recommend the use of calculators in schools, even in lower grades. This was part of an educational approach called "investigation." The students need not bother with tedious calculations. They should develop creative skills — "higher order thinking," was the slogan.

The mathematicians were aghast. Intuitively, they felt it was a crime. And their intuition was right. As already mentioned, calculation is not only finding the result. It is finding the decimal form of the result from the decimal form of the components. So, calculating entails a deep understanding of the decimal system. As mentioned above, the decimal system forms a large part of the elementary curriculum. A student who calculates 234×56 and understands what she is doing understands the decimal system.

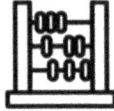

In the Country of One-Fingered Creatures

Imagine a country whose inhabitants have only one finger each. So, the idea of collecting tens, or even twos, never occurred to them.

They would represent the number 4 by

| | | |

and 13 as

| | | | | | | | | | | | |

The equality $9 + 4 = 13$ is then:

$$||||||||| + |\ |\ |\ |\ | = |||||||||||||$$

Is this a calculation? Not really, it is just the meaning of "9 + 4" — joining together the two groups, of 9 and 4. For one-fingered people calculation and meaning are the same. Math students there have a wonderful time! They only need to learn the meaning of the operations!

Before rushing to buy plane tickets to that country, write "1000" and "999" in their notation, and try to distinguish between the two. Despite all, we (pianists and harpists in particular) are lucky to have 10 fingers, or at least more than one. Some claim that it would have been even luckier if we had 12 fingers. 12 is a more convenient base, since in the "dodeca-decimal" system (gathering dozens) the rules for divisibility by 2, 3, 4, and 6 would be simple. To be divisible by 3, the last digit of the number must be divisible by 3, just as in the decimal system to be divisible by 5 the last digit must be divisible by 5, namely 0 or 5.

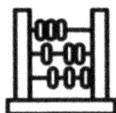

A Ten Is Born

When calculating $3 + 4$, we are just like the one-fingered people. Real calculation begins when you cross the border of 10.

The first stage is completing a 10. In class, pair the students, one in each pair should choose at most 10 pebbles (or matches) and put them on the desk, and the other should put on his side of the desk pebbles completing a 10. Then they should report: "I put 7 matches, my friend put 3." Did anybody put 3, to begin with? What did her partner put? Of course, 7. Indeed, $7 + 3 = 3 + 7$.

After practicing completion to a 10, it is time to calculate $9 +$ a number. Draw 9 circles, and on their side 5 circles. How many together? We can do it by what is called "continued counting" — after the 9, continue counting 5 steps, tracking the 5 on your fingers: 10, 11, 12, 13, 14.

But there is another way — let the 5 give circles to the 9 to complete a 10 — how many are needed? Just 1. There are 4 left, so we have $10 + 4$, which is 14.

Could the 9 give the 5 some circles to complete it to a 10? Yes, but this is harder — he must give 5, and we would have to know that $9 - 5 = 4$.

Practice this with $8 + 5$, $7 + 5$, $6 + 5$, $5 + 5$. Then ask what is $99 + 5$ — what if the 5 circles donate 1 to the 99?

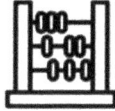

Why Are Math Notebooks Checkered?

Attending elementary school as an adult provided me with surprising new insights. Here are a few:

- Operations have meaning.
- Division has more than one meaning.
- What is a fraction?
- The decimal system is about (repeated) gathering objects in tens.
- Why are math notebooks checkered?

Why, indeed, do we use checkered notebooks? Because of the place value system. The squares in the notebook delineate the place of the digits. And even more importantly, the vertical lines help align numbers. When we add two numbers, it is important that the tens digit of one is above the tens digit of the other (and the same for hundreds, thousands…). The squares do the job.

A curiosity: to sum numbers, the Romans wrote them vertically, just as we do (they didn't have any reason to do so, they did not use place-value), and wrote the sum at the top — the "summit," hence the name "sum."

To start teaching vertical summation, write on the board:

$$
\begin{array}{r}
0 \\
+ \\
\hline
0 \\
\hline
0
\end{array}
$$

And

$$
\begin{array}{r}
2 \\
+ \\
3 \\
\hline
5
\end{array}
$$

The children will laugh, but they will appreciate this way of writing in:

$$
\begin{array}{r}
23 \\
+ \\
12 \\
\hline
35
\end{array}
$$

Here you see the reason — units above units, tens above tens. You can add them separately, units added to units, tens to tens.

As a child, I never inquired why addition, subtraction, and multiplication start from the right (the units), while division begins from the left. Here the answer is again simple — to avoid "regrets" — the need to retract operations. Here is an example:

$$
\begin{array}{r}
23 \\
+ \\
49 \\
\hline
\end{array}
$$

We start on the right: $3 + 9$, which is 12. Here comes the point: we can safely write 2 in the units' place of the result. Whatever we do next, will involve tens (or 100s if there are), not units. The 2 will not change. Had we started with the left digits, we would have gotten $2 + 4$, namely 6. But we could not safely write "6" in the tens place, because the 9 + 3 (in the units) contributes another 10. We would have to retract.

This is true for subtraction as well. Take

$$
\begin{array}{r}
72 \\
-\ 49 \\
\hline
\end{array}
$$

If we start on the left, we have 7 – 4 tens, which is 3 tens, but then there is 2 – 9, and we realize that keeping 3 tens was too optimistic, it will change to a 2. When we start on the right, we tackle 2 – 9 right away, and realize that we should unpack one of the 7 tens in 72, so as to get 72 = 60 + 12, and subtract the 9 from the 12, to get 3. Going to the tens, we now have 6 tens (one was used for the 12) minus 2 tens, which is 4 tens. Altogether — 43, and we did not have to retract any step.

The hazards of starting subtraction from the left.

Of course, in class, we do all this with objects.

Students who gathered tens with their own hands will easily understand that tens can also be dismantled, when necessary.

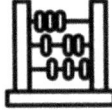

Should the Addition and Multiplication Tables Be Memorized?

Having a calculator on our desk, do we really need to know the multiplication table by heart?

Yes, we do. When a bricklayer builds a wall, the bricks should be ready in advance, not prepared as he goes along. Reading Harry Potter is not the right time to learn the letters of the alphabet. He who multiplies several digit numbers, had better master the building blocks of the algorithm, the multiplication of single digit numbers. When you learn how to calculate 47×44 you'd better have the result of 4×7 ready at hand.

Just as the multiplication table is the basis for calculating multiplication, the "addition table," namely the results of adding numbers smaller than 10, is the basis for calculating addition. It, too, should be memorized. It is easier than the multiplication table, and therefore not often discussed. But in the lower grades, it should be rehearsed until it becomes automatic.

The multiplication table is harder, but there are ways to make its memorization fun. First — counting by jumps. To learn the multiplication table of 6, jump forward 0, 6, 12,...,60 (don't forget 0×6 !) Then backward: 60, 54, 48,...,0. The multiples of 9 are easy — 7×9 is $10 \times 7 - 7$, so it is $70 - 7 = 63$. 5×8 is half of 10×8, namely half of 80, so it is 40. One of the hardest to memorize is 7×8 — use for it the mnemonic $56 = 7 \times 8$ (the numbers 5,6,7,8 appearing in order).

Another way to make memorization less dull is to find connections between products. For example,

$$10 \times 10 = 100, \qquad 11 \times 9 = 99$$
$$5 \times 5 = 25, \qquad 6 \times 4 = 24$$
$$3 \times 3 = 9, \qquad 4 \times 2 = 8$$

Do you see the pattern?

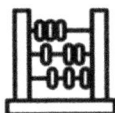

A Lesson on the Addition Table

One day I visited a Palestinian shepherd community in the West Bank. It was already dark (no electricity, of course), when a boy of 11 appeared from nowhere and sat next to me on a bench outside the tents. "Teach me math," he asked — he had heard I was a math professor. I asked him a few questions and realized he knew very little. "What is $6 + 6$?" I asked at some point. He didn't know, so I asked him to raise 6 fingers — naturally, he raised 5 in one hand, one in the other. I did the same. Now let us join the two. Without my prompting, the boy put his 5-raised fingers hand against mine: $5 + 5 = 10$. There remain the two 1-fingers hands — this gives 2 more. So, the answer is $10 + 2 = 12$. He was immensely pleased and asked to do more exercises of this type: for example, $7 + 8$. Again, two fives make a 10 in two hands, and 2 and 3 in the other two hands. It can be written as $7 + 8 = (5 + 2) + (5 + 3) = (5 + 5) + 2 + 3 = 10 + 5 = 15$. This trick gives all the addition tables, apart from sums having 1, 2, 3 or 4 as one summand, like $9 + 4$, but these are easy,

Teaching this in class starts with pairing up the students. Ask them to calculate this way all exercises with summands 5 or more: $5 + 5$, $5 + 6$,…, $6 + 6$, $6 + 7$…. Fun guaranteed.

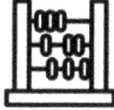

Vertical Multiplication

Amuse the children by writing 2×3 vertically:

$$
\begin{array}{r}
2 \\
\times \quad \\
3 \\
\hline
6
\end{array}
$$

What is the point? Bear with me. Do you remember how vertical writing was useful in calculating addition? It will be so also here.

$$
\begin{array}{r}
23 \\
\times \quad \\
2 \\
\hline
46
\end{array}
$$

Here the vertical writing helped: it separated the 3 units times 2, and the 2 tens times 2.

$$
\begin{array}{r}
23 \\
\times \quad \\
4 \\
\hline
92
\end{array}
$$

Here something interesting is already happening, a gathering of 10. You start from the right, 3×4, which is 12. There are 2 ones — you can safely write 2 in the ones place, everything else is going to contain tens. The ten you kept in your head is added to the 2×4 tens coming from the 2 tens in 23.

Next, what happens when the bottom number has more than one digit? Two simple examples:

$$
\begin{array}{r}
3 \\
\times \quad \\
10 \\
\hline
30
\end{array}
$$

and

$$
\begin{array}{r}
3 \\
\times \quad \\
100 \\
\hline
300
\end{array}
$$

Let us now jump into deep water: 24×332.

This is $(20 + 4) \times (300 + 30 + 2)$, which by distribution is $20 \times 300 + 20 \times 30 + 20 \times 2 + 4 \times 300 + 4 \times 30 + 4 \times 2$. Complicated? The wisdom of the generations taught us a short way of writing it:

$$
\begin{array}{r}
24 \\
\times \quad \\
332 \\
\hline
48 \\
720 \\
7200 \\
\hline
7968
\end{array}
$$

The first line, 48, is 24×2,
the second, 720 is 24×30
the third, 7200, is 24×300
All we need to do now is sum the three lines up!

Short Division

I already told you about the "investigation" revolution in math education. "Thinking, not drilling," "Thinking abilities before knowledge," "Principles, not details." The conclusion drawn was to renounce certain skills: fractions were to be learned only up to denominator 12, and the so-called "long division," the algorithm for division, was removed from the curriculum. It is too hard, was the claim. The students do not understand it.

But they can understand it, if it is demonstrated with concrete objects, namely with a "show." I have done it many times in Grade 3. The trick is to start with simple, even trivial examples. Remember? Nothing is too simple.

As usual, it is a good idea to start with something inane, say 6:1. Write it in long division form:

$$1\overline{)6} \;\; ^6$$

Next – by 2.

$$2\overline{)6} \;\; ^3$$

Why write it in this funny way? We shall soon understand. In the meantime, enjoy it.

Next step:

$$\frac{30}{2\overline{)60}}$$

Tens below tens — this is the principle behind vertical calculations. The 6 tens in the dividend and the 3 tens of the quotient are in the same column.

This should be accompanied by a show: a student holds 6 bundles of tens, each of ten sticks, and divides them into two parts, each consisting of 3 bundles.

Next:

$$\frac{34}{2\overline{)68}}$$

Again, demonstrate with popsicle sticks.

What if there is a remainder, as in 7:2?

Suppose you divide 7 apples between 2 children. Each gets 3 (sticks, please!). Have we divided all? No, only $2 \times 3 = 6$. Let us write it below the 7, and then $7 - 6$ is the number of sticks remaining, namely those we haven't divided as yet. One stick remains.

$$\begin{array}{r} 3 \\ 2\overline{)7} \\ -6 \\ \hline 1 \end{array}$$

This is the crux of "long division" — at each stage, we check what hasn't been divided yet. The "remainder." In this case, the remainder is 1. We are not going to divide it any further; it will remain a remainder.

Next:

$$
\begin{array}{r}
3 \\
2\overline{)78} \\
\underline{-6} \\
1
\end{array}
$$

In this case, the 7 is 7 tens. We divide them, the result is 3 tens, and the part we haven't divided is 1 — but in this case, it is 1 times ten. It is a ten that we haven't divided by 2 (and we should). What else haven't we divided? Of course, the 8. So we pull it down, next to the 1 (1 ten, we should remember, by keeping its position.)

$$
\begin{array}{r}
39 \\
2\overline{)78} \\
\underline{-6} \\
18
\end{array}
$$

So, we still have to divide 18. We compute 18:2 = 9. This we know from the multiplication table. We put the 9 in the place of the 1's, to get the final result: 39.

If you start from the right, namely from the 8 in 78, you get 8:2 = 4 in the units place, as a first stage. This is not right — we saw that the number of units in the result is 9. You will have to backtrack. Backtracking is the mire of calculation. Starting from the left avoids it.

This is the principle. Once you master this, it will be easy to continue further, to larger numbers.

Part 8

More Teaching Principles

Guessing

I asked my little daughter: how can you split 30 cookies between two children so that one gets 5 times as many as the other? She didn't know, so I asked her something that seemed easier — how to divide a cake between Bob and Carol so that Bob gets 5 times as much as Carol. She gave me right away the common erroneous answer — divide the cake in 5. Then she realized — if I give 4 parts to Bob and one to Carol, Bob will have 4 times as much as Carol, not 5 times.

This immediately led to the right solution — divide the cake into 6 parts, and give 5 to Bob, and one to Carol. Now it was easy to solve the question about 30 cookies — you divide into 6 parts, each of size 5, and give 5 of them to the first child — so he gets 25. The other child gets 5.

The guess was wrong, but it showed the way. Encourage the children to guess. Guessing enables sanity checks — confronting the solution with reality. It is also liberator from anxiety — guessing is not as committing as decisive answers.

Divide and Conquer

Many know how to juggle two balls. Juggling three is harder, but still achievable by amateurs. Four is for hard-working amateurs and five for professionals. The addition of every ball makes juggling many times harder.

So, should you start with two balls? No — better with one. This is one of the most basic principles of teaching. For example, division starts with division by 2, or even better — by 1. The children will be amused when you ask them to divide their 10 "flowers" between one person. In fractions you should start with "Egyptian fractions" — those having numerator 1, like $\frac{1}{2}, \frac{1}{3}, \ldots$ Once you do that, ask what is $\frac{1}{1}$, or even better, ask the students for the simplest Egyptian fraction. When you learn reading, start with "just one ball" — the consonants. Namely, start with a fixed vowel — ba, ga, da, ka, ta. cat, hat, sat, sad. Once the principle sinks in, move on to the next vowel — bet, set, met, jet.

We shall later see what happened in the US (and other countries following suit) when this principle was breached.

Approximately

Do not show a fool a half-done job.

— A Jewish proverb

A wise teacher asks half a question (and lets the student complete it).

— A teachers' proverb

Tamas Varga (1919–1987) was a famous Hungarian educationalist. When giving his students an exercise he would ask them first to guess (estimate) the result. What is 123×123? About 100×100, and those who know that $12 \times 12 = 144$ can have a better estimate: 14400. Asking for an estimate is "going halfway." Leaving fringes open arouses thought better than closed questions.

This practice is one of the best ways to develop a sense for numbers. And guessing should be extended to measurements (can you estimate the length of the class?) and weights (what is the weight of this apple, in grams? How many apples are there in one kilogram?)

Extreme Cases

Part of the tool kit of every mathematician is considering extreme cases. It should also be used in teaching — we already mentioned that teaching principles and thinking principles are not far apart.

Here is an example: Your salary was raised by 10%, and then reduced by 10%. Is it now higher than the original, or lower?

This is not very hard — the reduction was 10% of a greater quantity than the original salary, so the reduction was larger than the addition. You lose.

But there is a simpler way. Consider an extreme case — your salary was raised by 100%, and then reduced by 100% — is it now larger or smaller? Well, the second step reduced your salary to 0, it is now smaller. This indicates (though does not prove) that the same is true also in the non-extreme case.

Here is a more challenging problem. A circle of radius 1 rolls around a circle with radius 3. How many times does it rotate around itself? A concrete way of stating it: if a rod connects the center of the radius 1 circle with the circumference, how many full circles does it rotate when the small circle goes around the large?

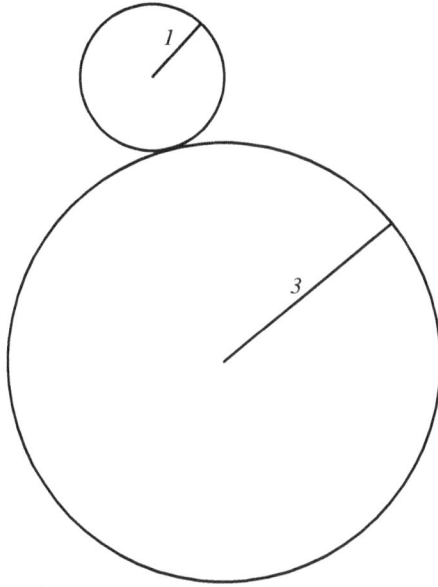

The small circle, having a third of the radius of the large circle, makes a full circle around the large circle. How many times does it rotate around itself?

Many answer "3." But this is wrong. If the small circle went along a straight line at a distance equal to the circumference of the big circle, it would indeed rotate 3 times. But it goes along a circle, which adds one rotation.

A good way to see this is to look at an extreme example: that the "inner" circle is of radius 0, namely a point. Obviously, if you go around a point, you rotate once. In this extreme case, the answer is 0 + 1, so the answer in the original question should be 3 + 1 = 4. This is indeed the case. Going around the "inner" circle adds one more twist.

"Do, Then Listen", or
"Listen, Then Do"?

Before receiving the ten commandments, the Israelites said — "We shall do, then hear," meaning that they would accept God's decrees without doubting, whatever they are. The Talmud, the commentary on the Torah, praises them for that. I am not sure I agree, but that's the Talmud's point of view.

Is this the right approach in teaching mathematics? Is it a good idea to first teach mechanically, and only then explain? For example, in the teaching of multiplication of fractions, is it wise to first teach the rule:

$$\frac{2}{3} \times \frac{4}{5} = \frac{2 \times 4}{3 \times 5}$$

(multiplying numerators separately, and denominators separately), then do more examples of this type so that the rule sinks in, and only then explain?

This is a mute question. My instinct is "No, the right way is to explain, step by step." Telling the rule with no explanation diminishes the motivation to understand. I personally never use the "first do, then listen" approach. But my own personal children demand it — "tell me the algorithm, then explain." Who knows, perhaps I shall be surprised once I try the other way. You can try both approaches and decide for yourselves.

The Archimedean Fulcrum
of Mathematical Education

Give me a leverage point and I will move the universe.

— Archimedes

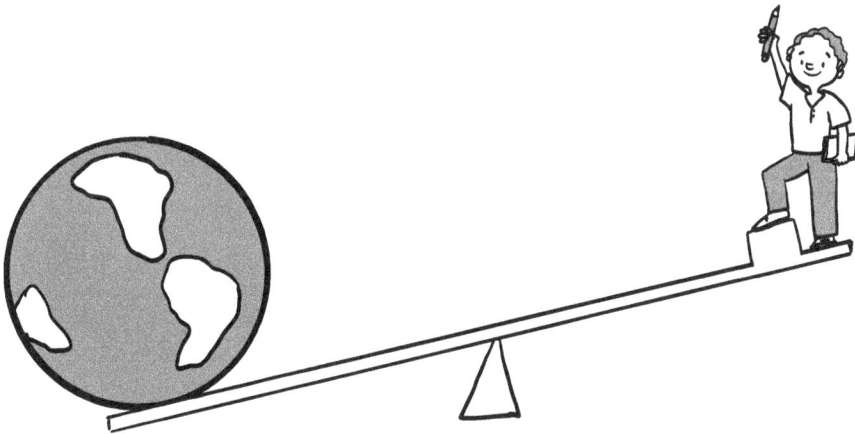

What determines the quality of education? The size of the classes, the quality of the textbooks, and didactic strategies. But research after research show that the critical factor is the quality of the teachers. In countries that excel in international exams, the status of the teachers, and their salaries, are high. In Finland, it is hard to obtain a post as a teacher, and even to enroll in seminars. There is fierce competition.

What can a poorer country, with different priorities, do? (Alas, there are such countries.) It sounds impossible, but in fact, there is an Archimedean fulcrum, in which a relatively small investment can yield a high return: in the teachers' education training. You can achieve a lot in four years of training.

Unfortunately, a large part of this time is wasted. A lot of redundant topics are taught in the seminar. Higher mathematics, cognitive psychology, and educational theories. The secret is simple: teachers teach what they know, not what the student needs. An expert on ancient Chinese will convince himself that this is what teachers need. There are no Chinese philologists in the programs for a Bachelor of Education, but there are mathematicians and experts in education and cognitive psychology. The expertise of all of these is not relevant for teachers. It would serve better to teach them the deeper points of elementary mathematics and practice teaching them.

I once tried to convince a teacher to have a class discussion. "This is not what they taught us in the seminar," she said. "I was taught that a class should consist of five minutes of exposure to a concept, and then individual work." The latter means solving problems from the textbook, each student at their own pace. The teacher goes from student to student and helps when needed.

What a wasteful strategy! One student at a time, instead of a whole class! More than that, the teacher herself would benefit from class discussion. It would force her to understand and to learn didactic techniques. Five minutes of exposure do not allow any deep diving. And it misses the most efficient teaching tool: class discussion.

Why have things gone this way? It is part of a general approach of educationalists — they assume that the teachers are not capable enough. The conclusion — we must teach over their heads, through the textbooks. This is always a mistake. How can the teachers learn their profession, when being skipped? The teachers' feeling of incompetence is a self-fulfilling prophecy.

Teach the teachers elementary math, the principles of teaching, and of thinking, and help them not to fear running a class discussion. It will enhance their abilities and their confidence. And best is that the teachers' teachers should be school teachers themselves.

Part 9

Fractions

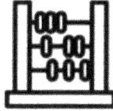

What Is $\frac{5}{8}$ of 240 — or, What Is a Fraction?

In the international exams of 1999, 75% of Grade 8 students in Israel didn't know what 5/8 of 240 is. In fact, it was a miracle that 25% knew, because nobody ever explained it to them. In Israel of that period, a fraction had a single meaning: a slice of pizza. And 240 is not a pizza.

Ask a 4th grader what a fraction is. Then go ask high school students, and then teachers. You will get the same blank faces, or non-informative answers, like "numerator and denominator" (what about them? What is the relationship between them?), or "part of a whole" (not every part is a fraction, and where have the numerator and denominator gone?). The right definition is never taught:

A fraction is the combination of two operations, division and multiplication, in this order.

This may look strange. A fraction is a number, not an operation! Well, by now we know that counting is also an operation, it is multiplication. "Half" is an operation, that takes half of an object.

So, $\frac{5}{8}$ of something is exactly as it sounds: 5 eighths. An eighth is obtained by dividing the object (the "whole") into 8 equal parts and taking one of them. 5 eighths are 5 such parts. $\frac{5}{8}$ of a whole is dividing the whole by 8 and multiplying it by 5.

Had the Israeli students been taught this definition, they would have known: divide 240 by 8, to get 30, and multiply by 5, to get 150.

Why do you combine division and multiplication, of all operations? Can't you do it also for subtraction and addition? Well, you could, but subtracting 8 from something and then adding 5 is just subtracting 3. The operation you get is not new, while a fraction is a genuinely new operation.

A fraction is like any number — it tells us "how many times you take the whole." It "counts" — so we have the answer to why is $\frac{1}{2}$ of 20 the same as $\frac{1}{2} \times 20$: just as taking 2 times an apple is 2 apples.

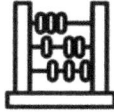

A "Must" Lesson

The following lesson is a must. I use it as an introductory class in Grade 3, as well as a corrective lesson in higher classes.

Please draw 10 objects to your liking: flowers, triangles, lines. Now circle a fifth of them — remember what a fifth is? You divide into five equal parts and take one of them.

I do it on the board together with the students. I write "a fifth of 10 is 2" and "a $\frac{1}{5}$ of 10 is 2."

Now circle another fifth. How many objects did you circle together?

I write "$\frac{2}{5}$ of 10 is 4." I continue this way, up to "5 fifths of 10 is 10." I ask them why, and a discussion ensues.

How much is 5 fifths of 100? Of one million?

They are proud of doing calculations in the millions. Next,

what is $\frac{6}{5}$ of 10?

Some are confused, but I remind them — one fifth was 2. How many are 6 fifths? We discuss the fact that you cannot circle more objects than are on the page, but you can think of it.

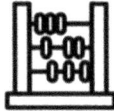

Fractions from Venus (Shapes), Division from Mars (Numbers)

Fractions should be introduced together with division, in Grade 2 or even in Grade 1. When teaching division by 2, introduce the notion of "half," including the notation — children love notation. Discuss: why is there 2 at the bottom of 1/2? Because there are 2 equal parts. Why 1 at the top? Because we are taking one of them.

In Grade 2 quarters and thirds should be introduced — in this order. "Quarter" is simpler, more easily drawn, and more useful (there is a quarter of a dollar, but no third, for example).

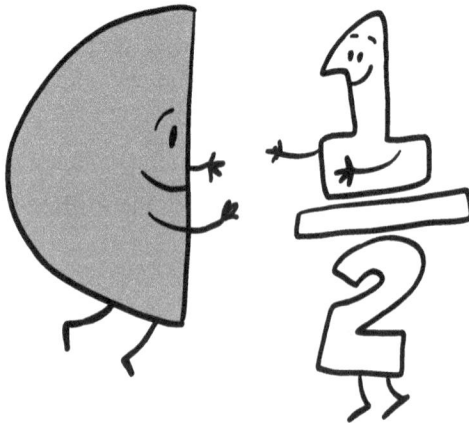

Why does the curriculum separate fractions from division? The reason is a mistaken conception: an assumption that division should be of numbers, while on first encounter fractions should be taken of shapes. "Fractions from Venus, division from Mars.[1]"

This makes no sense. Division should also be of shapes and bodies — divide a rectangle into two equal parts, an apple into 4, a round cake into 5, and a 30 cm string into 3. And fractions should be taken of numbers — I have 2 candies, what is half of them?

---

[1] A notion popularized by a book by John Gray (1992), bearing this name, arguing that men view relationships differently from women.

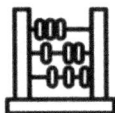

The Culprit: The Pizza Model

I was curious about the fractions-ignorance of the Israeli students — where did it originate? I looked at their textbooks. What I found was pizzas — the first 90 pages on fractions used pizza slices, not numbers. No wonder they did not know what $\frac{5}{8}$ of 240 is.

The popularity of the pizza model is no accident, it has a reason. Here is a rectangle.

Suppose I tell you that it is a fraction of a larger rectangle. Which fraction? And what is the whole rectangle? There is no way to tell. It could be a half, one-fifth, or even the entire, "whole" rectangle. With pizza slices, you know. Look at the shape below. You automatically assume this is part of a circle, hence it is a quarter of the whole.

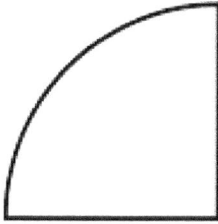

   I once gave three Grade 4 students a sheet of paper and asked them to give me a third of it. Their struggle was comical — for a long time they tried to "circle the rectangle," to draw slices emanating from the center. I had to tell them "this is a cake, please divide it among three children" — then they succeeded. Sticking to one model can be disastrous.

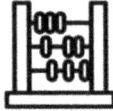

Juggling One Ball

A fraction has a numerator and a denominator. This immediately suggests applying "divide and conquer." First do what is called "Egyptian fractions," those with numerator 1, namely: $\frac{1}{2}, \frac{1}{3}, \frac{1}{4}$.... They are so named because the Egyptians had a special notation for them — an ellipse over the number of parts. For example, $\frac{1}{3}$ was written as:

3

An Egyptian fraction is just division — the whole divided (in this case) by 3.

As a first lesson draw a rectangle, a "cake," and ask a student to divide it between two children, evenly. They will probably draw

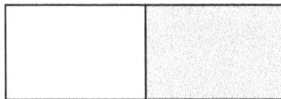

The way a cake is divided. Let them say it in words: "I divided into two equal parts, each part is a half." Encourage them to find other ways:

130

Draw

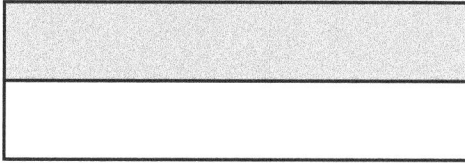

Can you now draw another possibility of dividing into two equal parts? Yes, there are many ways, here is one:

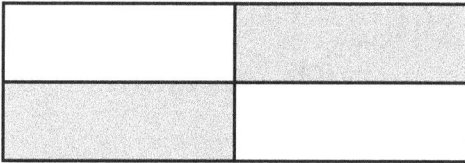

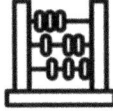

Teaching Fractions Together
with Division

There is no excuse: division should also be of shapes and objects, and fractions should be taken also of numbers. The students should divide a page into 3, 4, and 5, and a circle into 6 parts. It takes some abstraction to think of 24 as a "whole," but this is not beyond the comprehension of children.

> I am holding 24 flowers, what is half of them? And a quarter? Let us draw it on the board. Now draw 6 circles on your personal board and circle a third of them.

In Grade 2, I partitioned the board into 4 parts, by vertical lines. I summoned 4 children and asked another student to divide them evenly between the 4 parts. Of course, one to each. Then we did it with 8 children — of course, 2 in each part. Then I summoned 6 girls and asked them to divide evenly. Four of them went each to one part, and 2 girls remained. What should they do? They found a creative solution: bending so each had half her body in one part. We wrote $6:4 = 1\frac{1}{2}$.

After such an activity, ask the students to summarize the experience.

> ✍ A teaching principle: Once the class has performed some activity, summarize it right away, both aloud and in writing.

Why Is the Fraction Line a
Sign of Division?

Why $\frac{2}{3} = 2:3$?

No, it is not a matter of definition. On the left there is the fraction line, saying "two thirds." On the right, the result of division. Two apples divided between three children — what is each child's part?

If it is not a matter of definition, then it is a theorem. To prove it, think how you divide two equal rectangular pieces of cake between 3 people. Try this question on your child or students — they will probably find it hard. So, change the question: how do you divide a banana and an apple between 3 children? This is easy — divide the apple into 3 parts, and the banana into 3 parts. Now go back to the two rectangles — the children will do it easily — divide each into 3. Each of the 3 children receives one third from the first, one third from the second.

So, when dividing 2 cakes among 3 students, each gets two thirds. $2:3 = \frac{2}{3}$. The fraction line and division are the same.

In algebra, we prefer the fraction line over the sign of division. The reason is economy: the fraction line serves two functions. It is also parentheses, those that tell you which operation to do first. The rule is: you first calculate the numerator and the denominator (each separately) and then divide.

For example

$$(x+y):(x-y) = \frac{x+y}{x-y}$$

The fraction line serves as two pairs of parentheses — one sign instead of four.

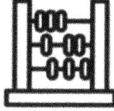

Multiplying Fractions

In numbers, addition is taught before multiplication. In fractions addition is hard, and multiplication and division are easier. They should be taught first. There is good reason for that — as we have seen, a fraction is the result of division, which goes along smoothly with multiplication and division.

Here are the steps:

1. Multiplying a fraction by a whole number: what is $3 \times \frac{4}{5}$? 3 times 4 apples are $3 \times 4 = 12$ apples. 3 times 4 fifths (think of a fifth as an apple) is 12 fifths. $3 \times \frac{4}{5} = \frac{3 \times 4}{5}$.

To multiply a fraction by a whole number you multiply the numerator by this number.

2. Dividing a fraction by a whole number: What is (4/5):3? In other words — what is $\frac{1}{3} \times \frac{4}{5}$? $\frac{4}{5}$ is 4 (say cakes) divided into 5 parts. If we divide instead by 3×5, namely into 15 parts, each part will be 3 times smaller. The more parts there are, the smaller each part. So,

$$\frac{1}{3} \times \frac{4}{5} = \frac{4}{3 \times 5}$$

To divide a fraction by a whole number, you multiply the denominator by that number.

135

Of course, this goes for every number, not only 3.

3. The general case — multiplying a fraction by a fraction: what is $\frac{2}{3} \times \frac{4}{5}$? $\frac{4}{5} = 4 : 5$, so multiplying by 4/5 is multiplying by 4 and dividing by 5. We know how to do that — multiply the numerator by 4 and the denominator by 5.

$$\frac{2}{3} \times \frac{4}{5} = \frac{2 \times 4}{3 \times 5} = \frac{8}{15}.$$

To multiply two fractions, multiply numerator by numerator, denominator by denominator.

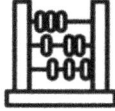

Dividing Fractions

In the chapter "The two meanings of division" we learned how to divide by 1/2. For example 10:(1/2) is "how many times 1/2 goes into 10", which is 20 times — every 1 contains 2 halves, so 10 ones contain 20 halves.

Conclusion: **Dividing by $\frac{1}{2}$ is multiplying by 2.**

Next, what is $30:\frac{3}{2}$? We know how many times $\frac{1}{2}$ goes into 30 – 30×2, namely 60. Instead, we now have "how many times does $\frac{3}{2}$ go into 30?" Each part is now 3 times larger, so there is room for 3 times less, namely 60:3, namely 20.

To divide by $\frac{3}{2}$ we multiplied by 2 and divided by 3. Which is the same as multiplying by $\frac{2}{3}$.

To divide by a fraction, multiply by its inverse, in which the roles of numerator and denominator are reversed.

Examples: $1:\dfrac{4}{5}=1\times\dfrac{5}{4}=\dfrac{5}{4}.$

$\dfrac{1}{8}:\dfrac{2}{8}=\dfrac{1}{8}\times\dfrac{8}{2}=\dfrac{1\times8}{8\times2}=\dfrac{8}{16}=\dfrac{1}{2}.$

Why is the last equality true? — we shall soon learn.

Famous lines of the English poet Alfred Lord Tennyson (1850–1892), on the pointless sacrifice of the English cavalry in the Crimean War (one of the most pointless wars in history in itself), go:

Theirs was not to reason why,
Theirs but to do and die.

American teachers, despaired of explaining the rule for division by a fraction, had a take on it:

Ours not to reason why,
just invert and multiply

Wars are hard to reason why, but in fractions it can and should be explained. Having done that using the arguments above, check the simplest examples: dividing by $\frac{2}{1}$ is multiplying by $\frac{1}{2}$, (that is, dividing by 2 is taking a half), dividing by $\frac{1}{2}$ is multiplying by $\frac{2}{1}$, (we already know that), dividing by $\frac{3}{3}$ is multiplying by $\frac{3}{3}$ (why?)

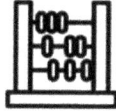

Expansion

Cut the pizza into only 4 slices, I am not hungry enough for 6.

— Yogi Berra, 1925–2015, a baseball player known for his tautologies

The fraction $\frac{4}{4}$ is not the same as $\frac{6}{6}$, they are written differently, but they have the same value — 4 quarters of a pizza or 6 sixths are both the entire pizza. We say that "they are equal, though not identical." Same size, though not the same form.

There is a simple way of obtaining equal pairs of fractions. It is called "expansion." If you multiply by a number and then divide by the same number, you return to where you started. This is what happened to Alice in the rabbit hole — shrinking and then growing back to the original size. Perhaps it has to do with the fact that Lewis Carroll, the author of "Alice in Wonderland," was a mathematician.

Now, we know how to multiply a fraction by (say) 5: multiply the numerator by 5. We know how to divide it by 5 — multiply the denominator by 5. So, if we do both, we get the original value. This is called "expansion by 5." For example, $\frac{1}{1} = \frac{1\times5}{1\times5} = \frac{5}{5}$, or $\frac{2}{3} = \frac{2\times5}{3\times5} = \frac{10}{15}$. Expanding $\frac{1}{2}$ by 3 gives $\frac{1}{2} = \frac{3}{6}$.

We can also go the other way round: divide both numerator and denominator by the same number. This is called "reduction." Reduction of $\frac{10}{15}$ by 5 gives:

$$\frac{10}{15} = \frac{10:5}{15:5} = \frac{2}{3}.$$

Reducing $\frac{100}{100}$ by 100 gives $\frac{1}{1}$, namely 1.

In the list below there are three fractions that have the same value. Can you find them? $\frac{1}{2}, \frac{3}{4}, \frac{30}{40}, \frac{6}{8}, \frac{5}{15}$.

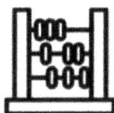

Addition and Subtraction:
The Common Denominator

3 apples plus 2 apples are 5 apples. Similarly, $\frac{3}{8} + \frac{2}{8}$ are 5 eighths. Saying it aloud clarifies the point. The two fractions share the same denominator. 3 apples plus 2 oranges is also doable: 5 fruits. There is a common denominator, "fruits." But what about $\frac{3}{8} + \frac{2}{6}$? To do it, we translate the two fractions. We write them both with the same denominator.

This is easy. If we expand $\frac{3}{8}$ by 6, and $\frac{2}{6}$ by 8, namely each by the other fraction's denominator, we get fractions having both 6×8, namely 48, as denominator. We have $\frac{3}{8} = \frac{3 \times 6}{8 \times 6} = \frac{18}{48}$ and $\frac{2}{6} = \frac{16}{48}$. So, $\frac{3}{8} + \frac{2}{6} = \frac{18}{48} + \frac{16}{48} = \frac{34}{48}$.

For the sake of elegance, we can reduce $\frac{34}{48}$ by 2, since both numerator and denominator are divisible by 2. So, $\frac{34}{48} = \frac{34:2}{48:2} = \frac{17}{24}$.

This means that in fact we could use 24 as a common denominator — it is divisible by both 8 and 6. $24 = 6 \times 4 = 3 \times 8$. So, $\frac{3}{8} = \frac{3 \times 3}{8 \times 3} = \frac{9}{24}$, and $\frac{2}{6} = \frac{2 \times 4}{6 \times 4} = \frac{8}{24}$, so our sum is $\frac{9}{24} + \frac{8}{24} = \frac{17}{24}$.

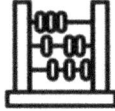

Smallest Common Denominator

It is not essential to find the smallest denominator, hence you can safely skip this chapter. But it is nicer. Here are two examples showing how to do it:

Example: $\frac{1}{15} + \frac{7}{12} = \frac{1}{3\times5} + \frac{7}{3\times4} = \frac{4}{3\times5\times4} + \frac{7\times5}{3\times5\times4} = \frac{39}{3\times5\times4} = \frac{13}{5\times4} = \frac{13}{20}$.

Example: $\frac{1}{15} + \frac{7}{54}$. Decompose: $15 = 3 \times 5$, $54 = 2 \times 3 \times 3 \times 3$. What number contains all factors that appear in both these two numbers, 15 and 54? $2 \times 3 \times 3 \times 3 \times 5$ does. It contains three 3's — you do not need a fourth, although 15 has another 3 in it: this 3 already appears in the product $2 \times 3 \times 3 \times 3 \times 5$. The rule is — take each factor the larger number of times it appears in the two denominators.

This means that $2 \times 3 \times 3 \times 3 \times 5 = 270$ is divisible by both 15 and 54. We can write

$$\frac{1}{15} + \frac{7}{54} = \frac{18}{270} + \frac{35}{270} = \frac{53}{270}.$$

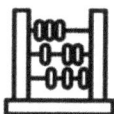

Comparing Fractions

An Israeli minister of culture taunted a contender: "You wish you had done a quarter of what I did. No — a third." You do not expect ministers to know fractions (in particular if they took their math lessons during the period of the "New Math"). But we do know: a third is larger than a quarter. When you divide a cake between 4 people, each gets less than when you divide between 3.

It is easy to compare fractions with the same denominator. Clearly, $\frac{5}{12} < \frac{7}{12}$. Things are also simple when numerators are equal — clearly $\frac{5}{12} < \frac{5}{11}$. But what happens if neither is the case? You use the common denominator.

Example: What is larger, $\frac{2}{7}$ or $\frac{4}{13}$?

One way to do it is to realize that $\frac{2}{7} = \frac{4}{14}$, which is smaller than $\frac{4}{13}$. But to do it systematically, use a common denominator — 7×13, which is 91. $\frac{2}{7} = 2 \times \frac{13}{91} = \frac{26}{91}$, while $\frac{4}{13} = 4 \times \frac{7}{91} = \frac{28}{91}$, so $\frac{4}{13}$ is larger.

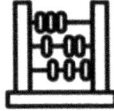

Mixed Numbers

Thou shalt not plough with an ox and a donkey together.

— Deuteronomy 22:10

The reason — they have different paces; they would suffer trying to synchronize. In mathematics it is allowed — you can mix integers and fractions, as in $3\frac{1}{7}$ (an approximation for the famous number pi, the ratio circumference/diameter in a circle). "Mixed numbers," as these are called, are ubiquitous. The trip from New York to Boston can take $4\frac{3}{4}$ hours, and you can rent an apartment of $3\frac{1}{2}$ rooms.

A mixed number can be converted to a fraction. For example, you know that $1\frac{1}{2} = \frac{3}{2}$. Also, $3\frac{1}{7} = 3 \times 7 + 1$ sevenths, namely $\frac{22}{7}$. Explanation: each unit contains 7 sevenths, so in 3 there are 3×7 sevenths. You probably see the rule for conversion, but to make it sink in, here is another example: to convert $10\frac{2}{3}$ to a fraction, you multiply 10 by 3, and add 2, to get 32: this is the numerator. So, $10\frac{2}{3} = \frac{32}{3}$. The numerator is larger than the denominator: such fractions are called "imaginary," to signify that they can be converted. There isn't really anything unreal about them. To convert an imaginary fraction to a mixed number, you just do the division: $\frac{32}{3} = 32 : 3 = 10(2)$. The (2) signifies a remainder of 2, that has also to be divided: it is really $\frac{2}{3}$. Altogether $10\frac{2}{3}$.

No new principles are needed here. In addition and subtraction, you just add and subtract the integer parts separately, and the fraction parts separately.

Example: $3\frac{4}{7} + 2\frac{1}{2} = (3 + 2) + \frac{4}{7} + \frac{1}{2} = 5 + \frac{8}{14} + \frac{7}{14} = 5\frac{15}{14} = 6\frac{1}{4}$.

In subtraction, we may have to convert a unit to a fraction. In the following example, we cannot deduct $\frac{1}{2}$ from $\frac{1}{7}$, because $\frac{1}{2}$ is bigger. So, we convert a unit in 3 to sevenths: $3\frac{1}{7}$ is $2 + 1 + \frac{1}{7} = 2 + \frac{8}{7}$, so:

$$3\frac{1}{7} - 2\frac{1}{2} = 2 + \frac{8}{7} - 2 - \frac{1}{2} = \frac{8}{7} - \frac{1}{2} = \frac{16}{14} - \frac{7}{14} = \frac{9}{14}.$$

Multiplication and division are computed by first converting the mixed number to a fraction. For example,

$3\frac{1}{7} = 3 \times 7 + 1$ sevenths, because in each unit there are 7 sevenths, so in 3 there are 3×7 sevenths. So, $3\frac{1}{7} = 3 \times 7 + 1$ sevenths, namely $\frac{22}{7}$. Likewise, $2\frac{1}{2} = \frac{5}{2}$, so $3\frac{1}{7} \times 2\frac{1}{2} = \frac{22}{7} \times \frac{5}{2} = \frac{110}{14} = \frac{55}{7}$.

You can discuss the similarity to the decimal system. Breaking 3 into 21 sevenths is similar to breaking a 100 to 10 tens, which is necessary, for example, in subtraction.

Part 10

Decimal Fractions

How do you know that an idea is clever? By its usefulness, and by the time it took to be discovered. Decimal fractions are immensely useful, and they appeared relatively late — only in the 16th century. In the form known to us today, they were formulated by Simon Stevin (1548–1620), a Flemish mathematician and physicist. We are so familiar with the decimal system, he said, why not extend it to the world of fractions?

How, precisely? In the number 324 the 3 counts hundreds, the 2 tens, the 4 counts ones. Going from left to right, each digit counts 10 times less than its predecessor. What happens if we add one more digit to the right, say in 3247? The 7 will count 10 times less than the ones (those counted by 4) — so it is $\frac{7}{10}$. Indeed? Not quite. We know well, in 3247 the seven counts 1s. The 7 pushed the number to the left — the 3 now counts thousands, not hundreds.

We need a barrier. A big, hefty wall, that will block the number from shifting left. But we are in the realm of abstraction, there is no need for a huge wall, a symbolic one, like "|," will do. And even better, just a point: 324.7. The point tells us that what is to its right starts a new bookkeeping: in 324.789 the 7 counts $\frac{1}{10}$'s, the 8 counts 10 times less, namely $\frac{1}{100}$'s, and the 9 counts $\frac{1}{1000}$'s.

The Singaporeans introduce decimal fractions as early as Grade 2, via money: for 4 dollars and 25 cents they write $4.25 — the previous familiarity with the concept saves energy.

Wonderful, isn't it? We now have our beloved decimal system also in the realm of fractions!

Well, there is no such thing as a free lunch. There are prices to pay, the first of which is the necessity of conversion between the two types of fractions.

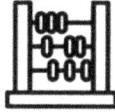

Conversion

Can you write 324.73 as a simple fraction? Easy: 324 and $\frac{7}{10}$ and $\frac{3}{100}$. Or simpler: $324\,\frac{73}{100}$, since $\frac{7}{10} = \frac{70}{100}$.

The conversion in the other way, from a simple fraction to a decimal, is harder. Worse than that — it is not always possible. You probably know at least one fraction that is not convertible to a decimal fraction: $\frac{1}{3}$. As is well known, it is 0.333.... That's not a decimal fraction. It is called an *infinite decimal fraction*. It goes forever. This means that the further you go before you stop (and you must stop, life is finite) you get nearer to $\frac{1}{3}$, but you never reach $\frac{1}{3}$ itself. Like Moses, who could only see the promised land from afar, from the east bank of the Jordan River, and was not allowed to get in.[1]

For a simple fraction to be convertible to a finite decimal fraction, its denominator, after reduction, should be divisible only by 5 and 2 and their products.

Example: $\frac{1}{500}$. The denominator, 500, is $5 \times 5 \times 5 \times 2 \times 2$: only 5s and 2s. Expand by 2 (to make the number of 5s the same as that of the 2s),

---

[1] It was a punishment for skepticism: God told him to speak to a rock to draw water from it, and to be on the safe side he also struck the rock with his cane.

and write $\frac{1}{500} = \frac{1 \times 2}{5 \times 5 \times 5 \times 2 \times 2 \times 2} = \frac{2}{1000}$ — the expansion gives 1000 in the denominator — a power of 10. So,

$$\frac{1}{500} = \frac{2}{1000} = 0.002,$$

tah-dah! — a decimal fraction.

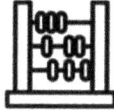

Infinite Decimal Fractions

This chapter is for those who would like to dive a little deeper.

Fractions whose denominator has a factor that is neither 2 nor 5 can be written as infinite decimal fractions. This is done plainly by division. Let us exemplify it using the fraction $\frac{5}{6}$. Spoiler: $\frac{5}{6} = 0.8333\ldots$

The denominator, 6, which is 2×3, has a factor, 3, that is neither 2 nor 5. Thus, as a decimal fraction, it is infinite.

We remember that $\frac{5}{6} = 5:6$, and we divide. Use long division, until the fraction starts to repeat.

$$
\begin{array}{r}
0.8 \\
6\overline{)5.0000} \\
^-4.8
\end{array}
$$

(4.8 is the part we have divided so far, 8 tenths times 6 — we have not divided all of 5.0. Subtracting 4.8 will tell us what remains to be divided.)

$5.0 - 4.8 = 0.20$ — this is the part of 5.0 that has **not** been divided so far.

Let us clean up:

$$
\begin{array}{r}
0.8 \\
6\overline{)5.0000} \\
-4.8
\end{array}
$$

Next we divide the 0.20 by 6, which is 20 hundredths divided by 6, which is 3 hundredths. We write it in the line of the result:

$$
\begin{array}{r}
0.83 \\
6\overline{)5.0000} \\
-4.8 \\
\hline
0.20
\end{array}
$$

Multiplying 0.03 by 6 tells us what we have really divided: 0.18.

Subtracting this number, 0.18, from 0.20, tells us what we haven't divided, so far.

This is 0.020, namely 20 thousandths.

We have to divide the 20 thousandths by 6 — this is 3 thousandths. We put 3 in the thousandths place of the result. But we divided this way only 18 thousandths — we must continue to subtract the part that remains.

$$
\begin{array}{r}
0.833 \\
6\overline{)5.0000} \\
-4.8 \\
\hline
0.20 \\
0.18 \\
\hline
0.020
\end{array}
$$

Etc. Once we get 2 as the remainder it will repeat. We get 0.833333… The "…" means "repeating forever."

Let us check if this is the right answer:

$$
0.8333\ldots = 0.8 + 0.0333\ldots = \frac{8}{10} + \frac{1}{10} \times 0.333\ldots
$$

$$
= \frac{8}{10} + \frac{1}{10} \times \frac{1}{3} = \frac{24}{30} + \frac{1}{30} = \frac{25}{30} = \frac{5}{6}
$$

Was it a coincidence that the remainders repeated, so that we got 2 again and again? No, there are only 6 possible remainders in dividing by 6, that is 0, 1, 2, 3, 4, 5 (actually, there are only 5 possible remainders: remainder 0 means the process stops, we divided all). So, after at most 6 steps the remainder repeats. And then the infinite fraction repeats from there on. In this case it was particularly quick — repeated from the second step and on.

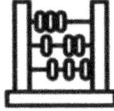

From Infinite Decimal Fractions
to Simple Fractions

How do you know that $0.333\ldots = \frac{1}{3}$? The secret is that $0.999\ldots = 1$. This means that the numbers 0, 0.9, 0.99, 0.999... get nearer and nearer to 1. The mathematical term is that they "tend to 1," or that "their limit is 1." The "limit" is the main notion in infinitesimal calculus, the branch of mathematics that studies minute changes, making it suitable for studying motion.

So, $1{:}3 = 0.999\ldots{:}3 = 0.333\ldots$ — each of the 9s is divided by 3. Nine tenths divided by 3 is 3 tenths, 9 hundredths divided by 3 is 3 hundredths, and so forth. All we are doing really is using the distributive law that tells us you can divide each summand separately.

This way, every repeating decimal fraction can be converted to a simple fraction. For example, to convert $0.123123123\ldots$ we write: $0.123123123 = \frac{0.123123123\ldots}{1} = \frac{0.123123123}{0.999999999\ldots} = \frac{123}{999}$, the last equality is true since for every 123 parts (like thousandths and millionths) that the numerator has, the denominator gets 999 such parts. So eventually the numerator is $\frac{123}{999}$ times smaller.

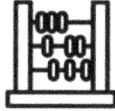

Addition and Subtraction
of Decimal Fractions

To add and subtract decimal fractions you arrange the numbers so that the decimal points are one over the other. To calculate 1.234 + 43.21 you write:

$$+\ \ \begin{array}{r} 01.234 \\ 43.210 \end{array}$$

I added a 0 at the beginning of the first number, and a 0 at the end of the second, just for typographical reasons, to make the location of the point clear. This way the 0 tens in the first is just above the 4 tens in the second, and adding them gives 4 tens. The same goes for ones (1 + 3 ones), tenths (2 + 2 of them), hundredths (3 + 1) and thousandths (4 + 0). So, you add the numbers as if they were ordinary numbers, just keep the decimal point in its place. In this example, the result is 44.444. The same goes for subtraction — ignore the decimal point, just remember to put it back at the end.

Popeye's Secret

In 1930, the conclusions of a research about the amount of iron in various types of food were published. 100 grams of spinach contain 3.8 milligrams of iron — nothing extraordinary, lots of foods beat it in this respect. But somebody missed the decimal point, and printed 38 milligrams — a hefty amount, 10 times the true value. This started the legend that spinach is rich with iron and started off Popeye's career.[2]

---

[2] Popeye is a cartoon figure, who uses spinach to get supernatural powers.

Moving the decimal point one place to the right multiplies the number by 10. For example, the number 234.56 becomes 2345.6. The two hundreds in this number become two thousands — ten times more. The three tens become 300, ten times more.

Of course, if you move the point to the left — in this case getting 23.456, the number is divided by 10.

This is at the base of the simplest way of multiplying decimal fractions. Here is an example: 2.3×4.56. Move the decimal point one place to the right in 2.3, and two places to the right in 4.56. You get 23×456 — an ordinary multiplication, that we know how to do — it is 10488. Moving one place to the right in 2.3 multiplied the number by 10, moving 2 places to the right in 4.56 multiplied it by 100, together the result was multiplied by $10 \times 100 = 1000$. We got 1000 times the true result, so to get the actual product we divide 10488 by 1000. We know how to do it — by moving the decimal point 3 places to the left: instead of 10488, which can be written as 10488.0, we get 10.488.

A sanity check: is it the right order of magnitude? Yes, 2.3×4.56 is close to $2 \times 4 = 8$, the result should be a bit larger than 8. It is.

Of course, the same trick works for division. Example: 2.31:1.1. Move the point two places to the right in 2.31, and one place to the right in 1.1, to get 231:11, which is 21. Now, moving the point two places to the right in 2.31 multiplied the result by 100, and moving it one place to the right in 1.1 **divided** the result by 10, since it made the divider larger by 10. So, altogether we multiplied the true result by 100:10, namely by 10. To get the true result, we must divide by 10, to get 21:10 = 2.1.

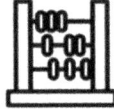

A Famous Fraction: Percent

For some reason, one fraction gained fame: $\frac{1}{100}$, called "one per cent" — one in a hundred. It is denoted by %, the slanted line reminding us that this is a fraction. Already the Romans gave it special status. For example, Augustus levied a tax of one per cent on auctions. Since then, we commonly think in percentages — nobody gets a salary raise of 1/12. Rather, you may get a raise of 8 or 9%.

Two factors combine to make this fraction useful. One is the affinity to the decimal system — as a decimal fraction, it is simple, 0.01. The other is that it often suits the dimensions of everyday life. Increments or loss of a few percent are common — in the GDP of a country, or in the size of the population. Getting 4% yearly interest on an investment makes sense.

Apart from that, there is nothing special about percentages. $\frac{1}{100}$ is a fraction like any other. Nevertheless, its prevalence in everyday life warrants some lingering. What is to follow is standard, but useful.

Conversion from percentages to fractions is easy: 50% is $\frac{50}{100}$, and a reduction (dividing the numerator and denominator by 50) gives $\frac{1}{2}$. The conversion is simply dividing the percent number by 100.

So, the other direction, conversion from fractions to percentages, is performed by the opposite operation, multiplying by 100.

Express the fraction $\frac{3}{250}$ as a percent.

$$\frac{3}{250} \times 100 = \frac{300}{250} = \frac{30}{25} = \frac{6}{5} = 1.2\%$$

A calculation that is often expressed in percentage is compound interest.

If you get 2% interest every year, what is the interest in two years?

2% is 0.02. So after adding 2% you have 1.02 of the original sum. The original sum was multiplied by 1.02. In the second year it is multiplied again by 1.02, so together it was multiplied by $\frac{102}{100} \times \frac{102}{100} = \frac{102 \times 102}{100 \times 100} = \frac{10404}{10000} = \frac{104.04}{100}$. So the compound interest is 4.04%.

Any question posed in terms of percentages can be solved by converting to fractions. For example, finding the whole by the part:

12% of which number is 60?

Answer: 12% is $\frac{12}{100}$. If $\frac{12}{100}$ of the number is 60, then $\frac{1}{100}$ of it is 12 times less, namely $\frac{60}{12} = 5$. If $\frac{1}{100}$ of the number is 5, the number is 100 times 5, namely 500.

Part 11

Educational Revolutions

Whole Words

What is the longest way between two points? — A shortcut.

Education is a strange realm — you can easily implement radical revolutions, especially if you are in the educational academy. Nobody will protest. The children are not informed, the parents trust the system and happily delegate responsibility to the school, and the teachers believe that academics are the carriers of divine wisdom. And the academics? Proud of their innovations, they are free to do whatever they want. The Ministry of Education is staffed by their friends — a Ph.D. in education is an entrance ticket there. That's how one of the major educational revolutions came about, unopposed — the change in the setting of the class. In the 1990s, the students were arranged in groups, seated around tables. This way half of them are seated with their backs to the teacher. All will be accepted, if just embellished by catchy slogans, like "creativity," "the student at the center," "freedom to think," "letting the children find their own way," and "active learning." Time and again educationalists draw a whole country (or even a hemisphere) to an unfounded experiment, whose damage takes decades to rectify.

Mathematics had its share, but the most harmful revolution was in the teaching of reading. For thousands of years, children (those who attended school) acquired the skill smoothly, following the rule "divide (between vowels and consonants) and conquer." It started with a fixed vowel — say "a": cat, hat, sat, that. The children understood the principle of the vowels, that the same vowel can be used with different consonants. Then they went on to the consonants.

This changed towards the end of the 19th century. One day in 1846 a rural teacher named John Russel Webb visited his neighbors, a woman and her young daughter. A cow was grazing in the yard, and he wrote the word "cow" on a piece of paper — "that's how you write 'cow'," he told the girl. She hurried to her mother, to tell her proudly — "I know how you write 'cow'."

Webb was so pleased, that he took the idea to his class. He wrote a few words on pieces of paper, and the children memorized them, and suddenly could skip the tedious bat-sat-cat routine. They learned whole words.

Later, in the 1920s, this was given an academic stamp. A psychologist named Gates advocated it — isn't it precisely the way toddlers learn to speak? Why should reading be different? The method quickly took over reading education in the entire United States. Opposition was derided:

> *The recent trend toward reinstating the purely mechanical word perception programs of the old alphabetic or phonic methods is viewed with alarm by educators who are interested in promoting growth in reading power.*
>
> — Paul Witty, phonic study and word analysis II,
> Elementary English, Vol 30

An influential proponent was the famous philosopher and education-alist John Dewey. In a paper entitled "The Primary Education Fetish" he advocated not teaching reading until the age of 8, and then doing it in "natural, non-intrusive ways."

Any sane teacher would have foreseen the ensuing catastrophe. But as usual, nobody consulted the teachers, the lowest on the pecking order of education. Instead of a few principles and 26 letters, the children had to learn a myriad of words. Most students just didn't manage it. A new profession was born — "corrective teaching," trying to help the weaker students, of course using the old phonetic approach — it was not the only time in the history of education that the weaker students got saner methods. This went on into the 1990s, when an extreme form of the "whole words" approach, called "whole language," wreaked such havoc on the teaching of reading that people, mainly parents, woke up and the method was ditched.

Similar revolutions occurred in math education, with the same detrimental effects.

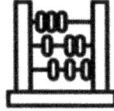

The New Math

It is not often that the teaching of mathematics makes its way to a popular song. The "New Math" did it. It was one of the strangest turns of events in the history of education. It drove students, parents, and mainly teachers, to despair. Tom Lehrer, a mathematician and a bard, wrote a deriding song about children who know that $3 + 2 = 2 + 3$ but don't know what the sum is. The idea is so unprincipled that its main interest is for sociologists of education.

In 1957, the Russians launched the first satellite, the Sputnik, into space. These were the days of the cold war, and panic gripped America — the Russians were ahead in science. The solution? Jump ahead! Study from the top, the abstractions. For example, Grade 4 students learned different base systems — replacing the decimal system with "binary" and "ternary" systems, where you gather sets of 2 and sets of 3 instead of 10, the base with which the students are familiar. Abstract, university-level math, made its way to elementary school. Teachers committed suicide.

The method spread to the entire Western world. For some fortuitous reasons in Israel more than anywhere else. Everything was taught via models, mostly invented by the creators of the system. Counting objects was discouraged and using fingers for counting and calculating was forbidden. Within a few years, the level of mathematical knowledge in the US dropped dramatically, while Israel went from first in the international exams to the bottom. In the US, the trend waned in the 1970s, in other countries it took longer — Israel parted with it only in the early 2000s.

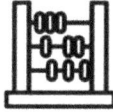

Investigation and the Math Wars

In the United States, the storm over the New Math subsided toward the end of the 1970s. But the dust hardly settled before a new revolution loomed on the horizon. We have already met it — "investigation," or "constructivism," signifying that the child should construct the knowledge on his own. No longer, said the proponents of this approach, should the child be a receptacle into which knowledge is poured. He should find things out by investigating the world. The teacher's role is that of a facilitator and supporter in the process of discovery.

This approach was implemented in the teaching of all areas of knowledge, but its effect was most significant in mathematics, because of its layered structure. Rigor was no longer assumed to be necessary, resulting in what the opponents of the method named "fuzzy math." For example, fractions were taught through cutting sections of circles, and as critics observed, in this representation children are easily led to believe that $\frac{1}{2} + \frac{2}{3} = 1$.

In 1989, the National Council of Teachers of Mathematics, NCTM, assembled the principles of the revolution in a book titled "The Standards." The book calls for redefining the goals of mathematical education. These are not knowledge, says the book, nor calculation skills. Material is not the essence. The goal is to achieve investigational abilities and deeper insights, for example, knowing how to offer conjectures or gather and process data. The words were exalting: creativity, the student in the center, teaching as a joint experience for teacher and student, and self-discovery.

The first US state to adopt the new approach was California. To the distress of Californian educators, the results were far from brilliant. They were as bad as the results of the "New Math." The main reason was the relinquishing of systematic teaching. Today we learn addition, tomorrow division, the next day subtraction, and then again division (division was very popular among the "investigationalists", because it offers diverse activities). Within five years, California dropped to 48th place among the American states in comparative mathematical tests. The percentage of students requiring preparatory courses in mathematics upon entering university increased by a factor of 2.5. Managers in California hi-tech companies realized that they had no local candidates to fill the available jobs.

When the reports of failure accumulated, a surprising thing happened: The exact science people awakened from their slumber. Many experienced the problem through their own children, guinea pigs of the system. Parents, mathematicians, and scientists united forces and declared war. This was the starting point of the "math wars," which have been stirring the American educational system ever since. In a petition to the Secretary of Education, published in the *Washington Post*, 220 well-known mathematicians and scientists called on the Secretary of Education to renounce the new approach.

In 1997, mathematicians were appointed to write a new, conservative California curriculum, and in 1998 it was introduced by law into all schools. It determines what the children need to know, and not how to teach it. Improvement was immediate. After some struggle, Israel followed suit and went back to sanity.

Part 12

Proportion

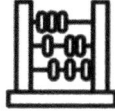

The Seven Sisters

Similarity

Visitors to Moscow are haunted by a strange apparition: wherever they go, the same building appears. You think you've gotten rid of it, and there it appears again in front of your eyes.

The University of Moscow, one of the Seven Sisters

The secret is that there are seven similar buildings strewn around the city. These architectural monsters are called "the Seven Sisters." They were built in the 1950s, not the result of frugality on design (although this, too, prevailed in Russia) but of communist megalomania. One of the Sisters, Moscow University, enjoyed the title of the "tallest building in Europe" for many years.

The Sisters are not identical but have similar proportions. This means that when you see a smaller one, you just assume it is further. We may see Aunt Miriam from close and from afar and still recognize she is the same person, although the image on the retina is different in the two cases. How do we do it? The two images have the same proportions between the parts of the face and the body. If in the close-up picture, the nose is twice as long as the mouth, it will be the same in the image of far-away Miriam. The height and the width of the Sisters are related (approximately) the same way in all. This is called "direct proportion." Our brain is an expert at identifying it.

Direct proportion is prevalent in real life. the cost of a basket of tomatoes is proportional to their weight. The amount of food we prepare is proportional to the number of people invited for dinner. The tax we pay is (sometimes) proportional to the income.

The Two Meanings of Direct Proportion

Direct proportion between two systems can be viewed in two ways.

A. Inner ratio: The ratios between parts of one system are identical to those of the other system.
B. Cross-proportionality: One system is identical to the other, times a constant number, say p. Each part in the second system is p times the size of the corresponding part in the first.

Let us exemplify it in the case of the Seven Sister buildings. Pick two Sisters, call them X and Y.

A. If in X the width is 2 times the height, then the same is true in Y.
B. All dimensions of X are (say) 3 times the corresponding dimensions in Y. If the height in Y is 100 meters, in X it is 300. If in Y the width is 300 meters, in X it is 900 meters.

Example: There is a direct proportion between the number of children in the class and the number of their hands. This can be expressed in two ways:

Inner: In every class the number of hands is twice the number of children.

Cross-proportion: If in one class there are 3 times as many students as in the other, it will also have 3 times as many hands (and 3 times as many ears, and 3 times as many noses).

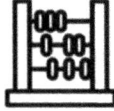

Per Unit: The Wisdom of km/Hour

The best-known case of direct proportion is that of speed. If a car drives at a constant speed, the distance it travels is proportional to the time it travels. Here comes a clever invention: measuring speed as distance per **one** hour. In general — per one unit. We measure speed by km/hour, or meters per second, or meters per hour in the case of snails.

A snail moves 3 meters in 2.5 hours. What distance will it cover in 4 hours?

To solve, calculate the distance it covers in one hour: $\frac{3}{2.5} = 3 : \frac{5}{2} = \frac{6}{5}$ meters. This is its speed. In 4 hours, it will travel $4 \times \frac{6}{5} = \frac{24}{5}$ meters, or $4\frac{4}{5} = 4.8$ meters.

This trick is useful in every problem on direct proportion.

4 shirts demand 5 meters of cloth. For how many shirts will 10 meters suffice?

"Per unit" is seen here in "how many meters per shirt," or "how many shirts per one meter." If 4 shirts demand 5 meters, then one shirt demands $\frac{5}{4}$ meters. Hence one meter is sufficient for $\frac{4}{5}$ shirts. 10 meters will suffice for 10 times more than one meter, namely $10 \times \frac{4}{5}$, which is **8** shirts.

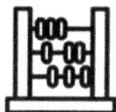

Inverse Rate, Pool Problems, and Parallel Resistors

A car travels the distance between NY city and Washington DC in 5 hours. What part of the way does it cover in one hour?

The answer, of course, is $\frac{1}{5}$ of the way. The original rate was hours/way, the inverse rate is way/hour. It is 1 divided by the hours/way rate (1/5 instead of 5). Easy? Indeed, but this is the secret to solving a class of problems dreaded by many — "filling a pool" problems.

Tap A fills a pool in 2 hours and tap B in 3 hours, how long will it take to fill the pool if both taps are turned on?

No, not 5 hours nor $2\frac{1}{2}$. It should be less than 2 — the 3 hours tap helps. The secret is to reverse units. Instead of "hour per pool" think of "pool per hour." Tap A fills half the pool in an hour, tap B fills a $\frac{1}{3}$ of the pool in an hour. So, together, in one hour, they fill $\frac{1}{2}+\frac{1}{3}$ of the pool, namely $\frac{5}{6}$ of the pool. To fill the entire pool, they need $\frac{6}{5}$ hours.

Just so you don't think there is something special about pools, here is another formulation:

Car A drives the distance between NY and Boston in 4 hours, car B in 5 hours. How much time will it take them to meet, if A leaves NY and B leaves Boston at the same time?

Instead of "hours per the NY-Boston distance" think of "distance per hour." Car A does $\frac{1}{4}$ of the distance in one hour, car B does $\frac{1}{5}$. So, when they join forces and drive toward each other, they drive $\frac{1}{4} + \frac{1}{5} = \frac{9}{20}$ of the way. To meet, they should drive together the whole way, which demands $\frac{20}{9}$ hours — a bit over two hours.

The formula is $\frac{1}{1/\text{rate } 1 + 1/\text{rate } 2}$ (in the last case - $\frac{1}{1/4 + 1/5}$).

For those of you who studied electricity, this formula may look familiar — it is the formula for calculating the compound resistance of two resistors that are connected in parallel. This is not surprising: the taps in the pool problem are "connected" in parallel — they are operated together, in parallel pipes.

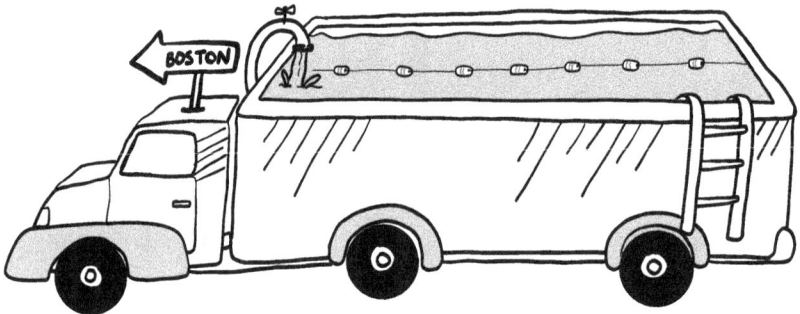

Part 13
Geometry

The Amazing Greeks

Four figures dominated Greek mathematics. Interestingly — none lived in mainland Greece. All four lived in the Mediterranean Greek colonies.

1. Pythagoras was born around 570 BC on the island of Samos in the Aegean Sea. He spent most of his adult life in Croton, a city at the foot of the Italian boot. There, he established a mathematical cult, which was eventually ransacked by neighboring cities. He probably died in that attack.
2. Euclid, who lived in Alexandria around 300 BC. Not much is known about his life. Some researchers even speculate that the name "Euclid" refers to more than one person.
3. Archimedes lived in Sicily 287–212 BC.
4. Diophantus, the great number theorist, lived in Alexandria between around 200 and 280 AD.

The name of Archimedes recurs in the book, and there is a chapter on the Pythagorean theorem. The spectacular achievements of Diophantus are too advanced for our book, so I will not mention him further. Euclid was responsible for developing geometry, the field containing the most impressive of the Greeks' achievements in mathematics.

One of Euclid's insights was the distinction between axioms, namely unprovable assumptions, and theorems, that can be deduced from the axioms. Another insight was the understanding that there is no need for drawings. "A point," and "a line" were for Euclid abstract entities. He observed five basic properties of points and lines and appointed them as the "postulates" (axioms) of geometry. The first of these is:

Given any two points, there is precisely one straight line passing through both.

Another formulation: through any two points passes a unique line. What is a point and what is a line? It does not matter, said Euclid. We have a common perception of what they are, and hence we know their properties. For example, between every two points on a line, there is yet another point. You can formulate these properties, without defining "line" and "point." Then you can prove theorems, without bothering to interpret these notions. When Russel said, "A mathematician doesn't know what he is speaking about," this is what he meant.

Are you ready? Brace yourself for a first proof — using (solely!) this axiom, we are going to prove the following theorem:

Theorem: Two distinct lines share at most one point.

Proof: Assume by way of contradiction that two lines, say K and L, meet at two points, say A and B. Then there are two distinct lines (K and L) that go through A and B. This contradicts the first axiom, by which through two points there is a single line. So, the negation assumption is impossible.

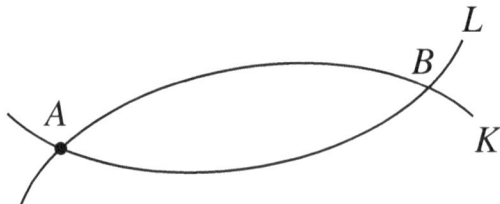

If lines K and L share two points, A and B, then there are two distinct lines going through A and B, contrary to the first axiom.

Congratulations on following a mathematical proof! It is an example of "proof by contradiction" — to prove the desired conclusion, we showed that its negation breeds an absurdity. Winnie the Pooh parodied this argument:

> *If anyone knows anything about anything, ... it's Owl who knows something about something, or my name's not Winnie-the-Pooh, which it is. So there you are.*
>
> — A.A. Milne, Winnie-the-Pooh

And look, Ma, no hands — no drawings! The drawing above guided our thoughts but was not essential. It is not necessary to know what a point and a line are to make this argument.

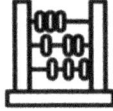

The Most Beautiful Proof?

What is the most beautiful mathematical proof? There are many contenders, and to answer we need to divide the proofs into categories. In the category of "short, by contradiction," there is an uncontested winner: Euclid's proof that there are infinitely many prime numbers.

A number larger than 1 is *prime* if it is not divisible by any number apart from 1 and itself. For example, 2 and 13 are primes, 91 is not, because it is divisible by 7 and 13 ($91 = 7 \times 13$).

Euclid's theorem is:

There are infinitely many primes.

There are infinitely
many primes

This is important because the primes are the building blocks of all numbers — every number is the product of primes. For example, $6300 = 2 \times 2 \times 3 \times 3 \times 5 \times 5 \times 7$. If there were only finitely many primes, number theory would be much poorer. This would be a shame, because its intricacy, vs. the simplicity of its notions, makes number theory beautiful, many say — the most beautiful mathematical branch. In no other field can you find such simple statements that are so hard to prove. An example: "There are infinitely many pairs of twin primes" (that is, differing by 2, like $(3,5)$, $(11,13)$, $(101,103)$). This conjecture is still open, after centuries of determined attacks.

But let us go back to Euclid's proof. It requires two preliminaries:

I.   Every number n larger than 1 is divisible by some prime.

For example, 15 is divisible by 3 (also by 5, but we only need one prime divisor). Why is this true? If n itself is a prime, then we are done — it divides itself. If not, then n is divisible by a smaller number, say n_1. If n_1 is a prime, we are done — n is divisible by it and it is a prime. If not, then by the definition of primality, n_1 is divisible by a smaller number, say n_2. Since n_2 divides n_1 and n_1 divides n, also n_2 divides n. If n_2 is a prime, we are done. If not then it is divisible by a smaller number n_3, and so on. The process must stop because $n > n_1 > n_2 > n_3...$, and there is no infinite sequence of descending natural numbers (can you explain why the sequence above stops after at most n steps? Hint: at each step, the number decreases by at least 1.) When it stops, we have reached a prime number n_i that divides n.

II. If p divides m, then p does not divide $m + 1$. For example, 3 divides 300, so it does not divide 301, there is a remainder of 1.

Now to Euclid's proof that there are infinitely many primes. It is done by contradiction. Namely, showing that the negation leads to absurdity.

Suppose (for the sake of contradiction) that there are only finitely many primes (say 2,3,5,7 and 11 are all the primes in the world). Let m be their product — the result of multiplying all of them (here we use the assumption that there are only finitely many primes. We know how to multiply finitely many numbers. In our example $m = 2 \times 3 \times 5 \times 7 \times 11 = 2310$). Obviously, m is divisible by all primes (the assumption is that all primes were included in the product). So, by item 2 above, $m + 1$ is not divisible by any prime. But this contradicts I, by which every number is divisible by some prime. In our example $m + 1 = 2311$ is itself a prime, contradicting the assumption that 2,3,5,7 and 11 are all primes in the world.

An Axiom, or a Theorem?
The Story of the Fifth Axiom

For 2400 years, everybody believed that Euclid's five axioms capture all there is to know about plane geometry. At the end of the 19th century, somebody more critical appeared on the scene: David Hilbert, the great mathematician of that period. He realized that the five axioms conceal many implicit assumptions. He formulated twenty such assumptions and dubbed them "axioms." For example, there is an axiom that says "between any two points on a line there exists a third point lying on that line."

> Challenge — can you use this axiom to prove that between any two points on a line there are infinitely many points?

Hilbert thereby answered in the negative the question "Do the five axioms of Euclid compound all the truth?" But for 2200 years another problem tormented the mathematical world: are all five really axioms? Isn't it possible that at least one of them can be proved from the others? In that case, it would not deserve the title of "axiom." It would become a theorem. The main suspect was the "fifth postulate" (a spoiler — it turned out to indeed be an axiom, independent of the other axioms). It is this:

For every line *L* and a point *P* not on it there exists precisely one line passing through P and parallel to *L*.

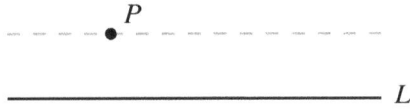

This was called "the parallels axiom." For more than 2000 years, mathematicians tried to prove it from the other axioms and failed. The struggle ended around 1830, when a young Hungarian mathematician, Janos Bolyai (pronounced "Boyai") showed, using an ingenious argument, that it is indeed an axiom — it does not follow from the other axioms. He did it by constructing a world in which the other axioms are true, and the "parallels axiom" is false. If the fifth postulate logically followed from the other axioms, then in any world in which the other axioms are true the fifth postulate would also be true. And this is not the case, so showed Bolyai.

Bolyai's personal story was less happy. His father sent the discovery to his friend Gauss, and Gauss answered: "I cannot praise your son's discovery, because it would mean praising my own." In other words, I proved it long ago, sorry. Or not so sorry. Gauss hardly ever published his discoveries. His claim was not totally justified — Bolyai took the discovery further and understood more. Heartbroken, he practically quit mathematics.

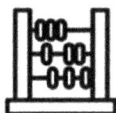

Main Points

Geometry is the study of shapes: points, straight lines, and circles. About a quarter of the curriculum is devoted to it. A well-deserved amount, for two reasons.

1. Geometry is the abstraction of something essential in our lives — location in space (or, first, in the plane).
2. Geometry is an arena for a first encounter with mathematical proofs.

And there is a third reason — beauty. Geometry is strewn with gems, that even children can appreciate.

For these advantages, geometry pays with a lack of structure, at least as compared with arithmetic. The curriculum in arithmetic can be summarized succinctly: the operations, and the decimal system. Elementary school geometry is less compact and coherent. It is too rich to summarize in a short text like this book. It has a multitude of concepts and elementary facts. I therefore chose to concentrate on just a few subjects.

1. A first encounter with the basic notions. How to make the children experience them.
2. Examples of "deduction" — namely proofs.

(3) The idea of "geometric location" — the set of points satisfying a specific property. Like the circle, which is the set of points having some specific distance from a point called the "center."
(4) Similarity.
(5) The circle and the ball (yes — we shall also go into three dimensions).

Some of the proofs are beyond the abstraction powers of most elementary school students. But we don't need to go into intricate proofs. The main point in elementary school geometry is drawing — and guessing facts. An example: a triangle with all sides equal has equal angles. In elementary school, we don't have to prove this. The important part is letting the students guess it.

Geometry vs. Number Theory:
Which Is More Complex?

Let me pay number theory its deserved tribute. Elementary geometry is much richer than elementary number theory. It uses more concepts, has more theorems, and in general, looks more varied. In higher mathematics the opposite is true. Number theory is infinitely more complex. Geometry can be mastered by a computer: there is a computer program that examines every geometric statement and decides whether or not it is true. If it is, the program can provide a proof. A famous "incompleteness theorem" of Kurt Gödel from 1931 says that there will never be such a program for number theory. There cannot be a program settling the validity of statements in number theory. No program will tell you about every statement it receives whether it is true or not. Number theory is too complex.

First Lesson: Points and Lines

Georges Seurat (1859–1891) was a great promise in French art. He didn't completely fulfill the promise, because he died young, at 31, and because he chose a very slow form of painting — point by point. The painting below, "Sunday afternoon," took him two years to complete.

Let the students be little Seurat's

Let us be little Seurat's. Ask Grade 1 students to draw a house using points. Can you count the points you used? Now draw a house using only straight lines. Can you draw a cat using only straight lines?

Next ask the children to draw two points on one page, that are far from each other. Then two points that are close. Then a point to the left of another, and a point above another. Draw their attention to the fact that the "higher" point is not really higher — they are both on the paper, that is on the table! Why do we call one "higher"? To ease the answer, let them draw a person — where is his or her head, and where are the feet?

The next basic notion is that of "line," short for "straight line." Ask the children to draw a straight line in their books, or on their personal boards. Then, look for straight lines in the room. Move your hand along the rim of the table, and feel that it is straight. Move your hand along the rim of the window. Geometry lessons are an opportunity for moving and for fun — do them at the end of the day.

Draw a person, and then draw a line connecting his or her ears. This line is called "horizontal." Now a line from his or her head to the ground. This line is "vertical."

With your hands, draw an imaginary horizontal line, and then an imaginary vertical line.

Now ask the children to draw two lines. With a little bit of luck, somebody will draw two lines that meet. Ask them to draw two lines that do not meet, but extended beyond the page, they do.

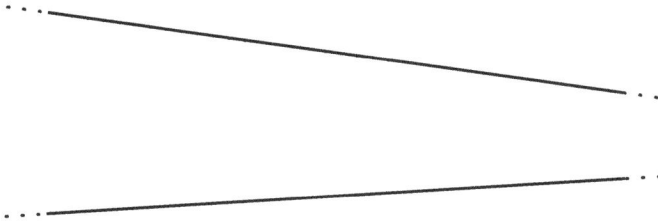

These lines do not meet on the page, but their extensions do!

Finally, ask them to draw two lines that will not meet even when extended. These are called "parallel lines." Do you see pairs of parallel lines in the room? Of course, plenty. Two opposing rims of the table, two opposing sides of the window, two opposing sides of a page.

Put your two hands forward — are they parallel? Ask the children to put their hands forward in a parallel way, and non-parallel.

Draw a pair of parallel segments, one longer than the other — are they parallel? Many will say "no, parallel lines have equal length." Extend them, to realize they are parallel.

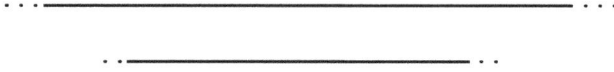

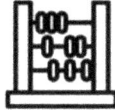

Definable vs. Undefinable

What is a point? What is a line? These cannot be defined in words (try!) They must be drawn and experienced. Other concepts can be defined. This is a good opportunity to discuss definitions. What are they? With some help, the students will come up with something like "a definition is a description that tells you precisely what you are speaking about."

You can spend enjoyable hours discussing definitions. Enjoyable and productive — definitions make you think. Can you define a "dog"? "An animal with four legs that barks." Aren't there dogs that do not bark? Yes, there are. "Domesticated animal from the family of the wolves" — not bad. But what is a wolf?

In higher grades, you can point out to the students that there are much harder concepts. What is "I"? What is a "thought"? But let us move on to things we can define. For example, "A triangle is a figure consisting of three points, and the three segments connecting them."

Here are four basic concepts that should be defined. Draw and ask the students for definitions. A worthwhile exercise.

A **Segment** is a part of a line between two points, that are called the "endpoints" of the segment.

A **Polygon** is a closed shape, constructed by appending segments, each in turn. The segments are called "edges," and the meeting points of consecutive edges are called "vertices."

An **n-gon** is a polygon with n sides (this implies having also n vertices).

A **ray** is "half" a line, the part of a line from one point to one side of it, all the way to infinity. We call the point the "apex" of the ray.

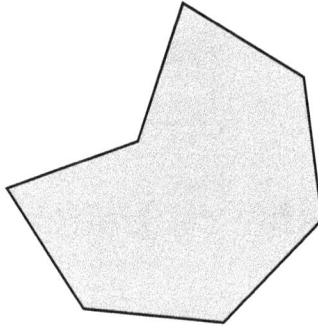

A 7-gon

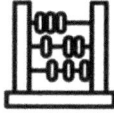

Angles

The next notion, "angle," is harder to define. An angle consists of two rays emanating from one point, but its essence is in the aperture of the two rays. The closest I (and others) can come up with is "the area enclosed between two rays sharing the same apex."

Here is an acute angle:

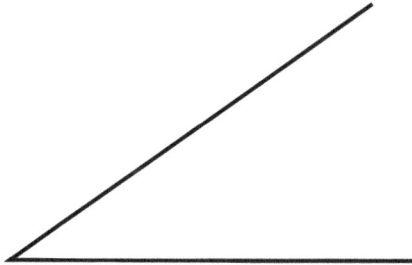

It can stab, "acute" means sharp. Can you form a sharp angle with your hands? Make it even sharper! Now make it more obtuse, namely less sharp. Draw on paper the most acute angle you can. The students will compete, but possibly none will dare to draw an angle of size 0. You can discuss this possibility — even use the term "degenerate" — it is a degenerate angle, in which the two rays coincide. It is knife-sharp.

What is the largest angle? The students will probably suggest a flat angle — also called "straight." In fact, there are larger:

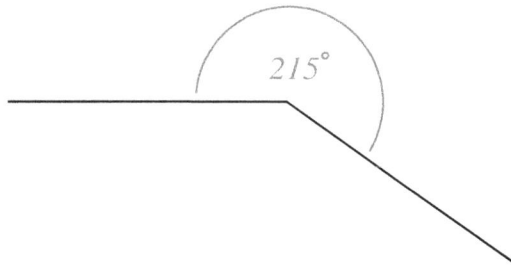

In Grade 3, we can tell the students that we divide the straight angle into 180 parts, that are called "degrees." Why 180? No reason. It is just a nice number, divisible by many numbers — 2, 3, 4, 5, 6, 9, 10, 12, 15, 20, 30, 60. The idea of dividing into an *arbitrary* number of parts is abstract. Ask the students why the meter is chosen as it is: no reason. It is arbitrary. The same goes for dividing the flat angle into 180 degrees.

Next — the right angle, that is half of the straight angle. How many degrees are in it? Of course, 90. And how about the "knife," the two coinciding rays? Indeed, it has zero degrees.

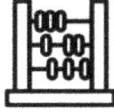

The Right Angle

The right angle is worth spending a few lessons on. Here is one:

It is half of the straight angle: two straight angles make 180°, which is a straight line. Summon a student to the front of the class and ask them to walk to the wall facing them in the shortest way possible. Do it also in drawing — let every student draw in their notebook, or better — on their personal board, a point and a line not containing it. The "point" in the picture represents the student facing the wall. What is the shortest way to the line (the wall)? It forms a straight angle with the wall.

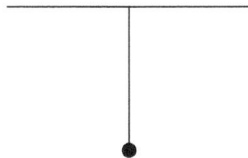

Look around — do you see right angles? Of course, a bounty. At the corner of the table, the corner of the page. The legs of the table form a straight angle with the ground.

Why are there so many right angles in the world? What would happen to a wall that does not form a right angle with the ground? It may fall. What would happen to the rooms in your apartment if they did not have right angles? They would not fit together.

Use the word "perpendicular." The walls of the class are (presumably) perpendicular to each other. Other pairs? Of course — the sides of the pages.

An angle larger than a right angle is called "obtuse." Can you form an obtuse angle between your hands? Draw an obtuse angle. If you continue one of its rays, what angle do you get adjacent to the obtuse angle? It is called "the complementary angle." Complete the sentence: "The complementary angle to an obtuse angle is ....."

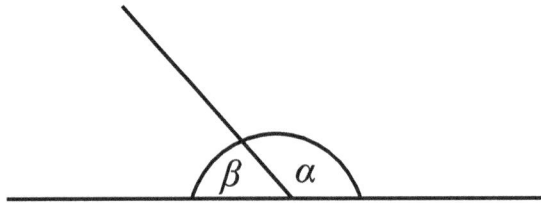

An obtuse angle α and its acute complement β

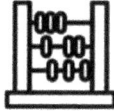

Opposite Angles

One of our university professors told us that a proof is beautiful if there is a new element in it, that doesn't appear in the problem and yet is instrumental in reaching the conclusion. The next proof satisfies this condition.

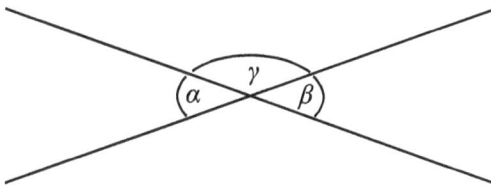

Opposite angles

Our eye tells us that in the drawing $\alpha = \beta$. In other words, "opposite angles are equal."

But why? The angle γ plays the role of "deus ex machina."

Look at the drawing — α and γ complete a straight line together. So $\alpha + \gamma$ is a straight angle, and so is $\beta + \gamma$. So, $\alpha + \gamma = \beta + \gamma$.

This can only be true if $\alpha = \beta$.

As a child, I was impressed. I still am.

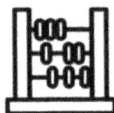

More on Parallel Lines: Infinity and Transitivity

It is time for a definition of "parallel lines."

Two lines are parallel if they do not meet, even when extended.

Draw a line on the board, and ask — can you draw a line parallel to it? And another? How many lines are parallel to it? Indeed, infinitely many. Do you know anything else of which there are infinitely many? Indeed, numbers. Something else?

Here is a line. Draw a point on it. And another, and another. How many points are there on the line?

Is there anything in real life that is infinite? Is the whole world (that is, the universe) infinite? You don't see an end, but this is no proof of infinity. If you are in a circle, you don't see an end either, but this doesn't mean that the circle is infinite! An ant on a ball thinks it is infinite — it does not see an end.

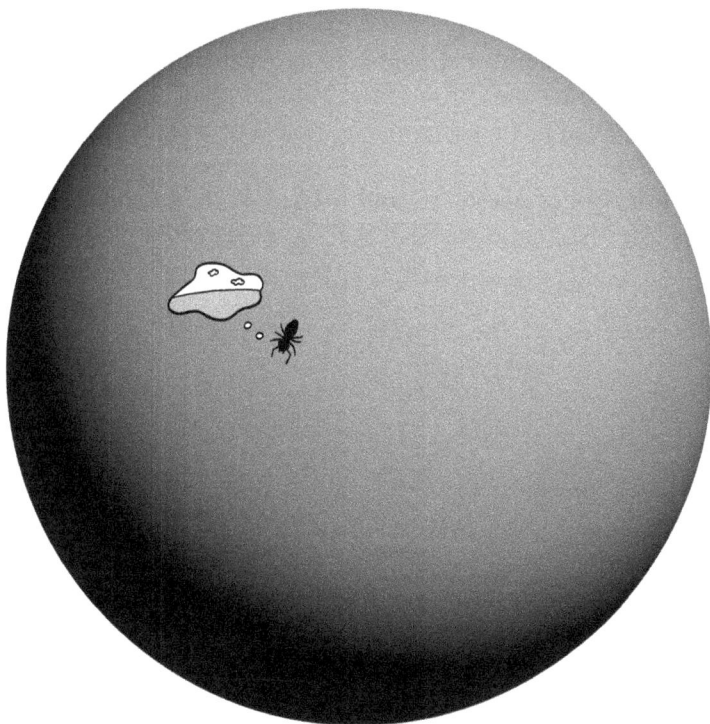

The ant thinks the ball is infinite and flat

Draw a line, then a parallel to it, and then a parallel to the second line — is the last line parallel to the first line?

Can you formulate the rule? The students may falter, something like:

If a line is parallel to another line, and another line is parallel to it...

The tongue stumbles. Offer them to name the lines. Without names it is like speaking about people without naming them, using only "He, she...."

If line B is parallel to line A, and line C is parallel to line B, then line C is parallel to line A.

C ····——————————————————————————····

B ····——————————————————————————····

A ····——————————————————————————····

Transitivity: if line B is parallel to A, and C is parallel to B, then C is parallel to A.

This is called "transitivity." Let us try it with another relation: if number B is larger than number A, and C is larger than B, is it true that C is larger than A? (Of course). How about "smaller than"?

Transitivity can be understood by Grade 5 or 6 students. Write on the board:

If person B likes person A, and C likes B, then C likes A.

Is this true? Not necessarily. "Like" is not transitive.

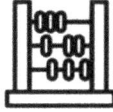

Corresponding and Alternate Angles

When a line crosses two parallel lines eight angles are formed. Can you point them out? Three are marked in the following picture:

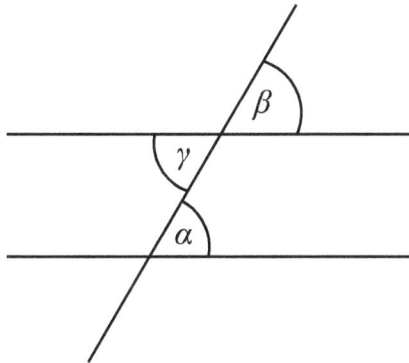

α and β are called "corresponding," and α and γ are "alternate."

The eye tells us that all three are equal, and indeed they are. Why? How do you prove that $\alpha = \beta$? Well, you don't. Not because it is an axiom, but because it is a definition. Not only axioms are not proved — but also definitions. In this case — the definition of equality between angles. The definition includes as a special case the equality of corresponding angles. Also, in the picture $\beta = \gamma$ because they are opposite. So, $\alpha = \gamma$. α and γ are called "alternate angles."

Part 14

Congruence and Similarity

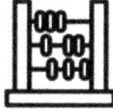

The Triangle Inequality

There is a famous inequality concerning triangles: the sum of the lengths of two sides is greater or equal to the length of the third side.

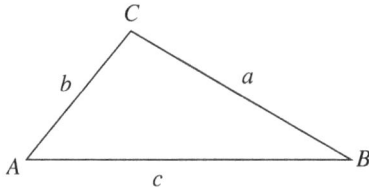

The triangle inequality says that $b + a \geq c$

In fact, I am sure you already know it: it is just the well-known fact that a straight line is the shortest path between two points. c is the straight line distance between A and B, while $b + a$ is a circuitous way, via point C.

Congruent Triangles

Two shapes are called "congruent" if they are "equal." But what does it mean, "equal"? It means that you can put one on the other with perfect fitting. "Put" means moving and possibly reflecting (as if in a mirror), without changing the inner structure. This is the formal definition. In class it suffices to define it by cutting one shape (start with a triangle) out of the paper and putting it over the other shape. "Congruence" means the two shapes fit perfectly. Show the students triangles that need flipping (reflecting) to see they are congruent.

How much information do you need about a triangle to construct it? Ask the students how many pieces of information are needed to describe a person uniquely. Will his first name suffice? Of course not, there are many "John"s. Adding the last name? Better, but may not be enough. Perhaps also his city of residence? Adding the precise address will probably suffice.

Let us now move to geometry. Suppose you have a triangle drawn in your notebook, and you want your friend to draw in her notebook precisely the same triangle — how many pieces of information do you need to tell her? Will the length of one side suffice? Of course not, there are many different triangles with this side. Two sides? No, they can have different angles between them.

Three sides? Yes, three sides suffice. If you know the lengths of the three sides of a triangle, you know the triangle. Try to draw and convince yourself. This is called the "SSS (side-side-side) congruence theorem":

SSS: Two triangles with the same side lengths are congruent.

Let the children experiment to convince themselves this is true.

How do you draw a triangle *ABC* knowing the lengths of *AB*, *BC*, and *AC*? First, draw *AB*, its length is given, so this is easy. Then draw two circles — one centered at *A* and having radius *AC* and the other centered at B and having radius *BC*. The circles may meet at two points, choose one of them to be *C*.

It is possible that the circles do not meet — take as an extreme example the case that the lengths of both *AC* and *BC* are 0. In this peculiar case the two circles are just points. Then there is no triangle with the given lengths. Reason: the lengths violate the triangle inequality.

The two other theorems are easier to prove:

SAS (side-angle-side): Two sides, and the angle between them, determine the triangle.

To see this draw two rays forming the given angle, and mark the lengths of the two sides on the two rays.

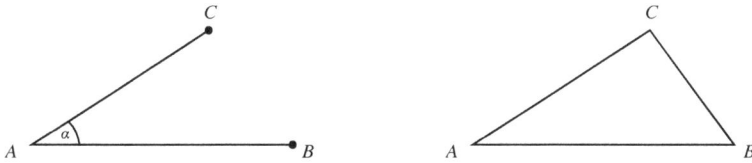

On the left the given angle and lengths of sides, on the right the triangle

ASA: A side and the two angles next to it determine the triangle.

The drawing is almost self-explanatory: just draw the angles on the ends of the given side and prolong the rays until they meet.

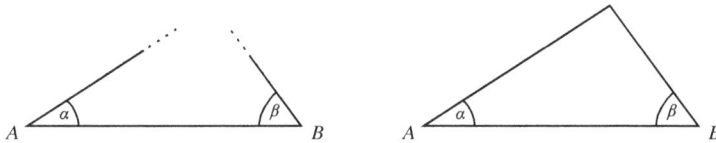

ASA: On the left the angles are drawn, on the right the rays meet

This is called the "ASA" (angle-side-angle) theorem.

So, three pieces of information (sometimes) suffice. How about the three angles? They do not suffice. Two equilateral triangles have the same angles (all angles are 60°), and they may be of different sizes. But of course — three angles are not really three pieces of information — they are really two. Two angles determine the third, because, as we shall soon prove, the sum of the angles in a triangle is 180°.

Isosceles Triangles

A triangle with two sides having the same length is called "isosceles." The third side is then called a "base."

Theorem: The angles near the base of an isosceles triangle are equal.

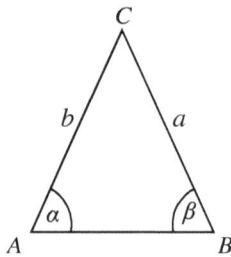

The reason — the triangle ABC in the picture is congruent to ACB, the triangle obtained by flipping. (As a child, I was suspicious of this argument — but it is totally valid). So, the corresponding angles are the same.

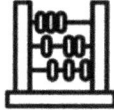

Parallels and Similarity

Two shapes are called "similar" if one of them is obtained by magnifying the other.

Two similar cats — the right one is a 3-fold magnifying of the left one.

The image on the retina of a cat and the image of the same cat, viewed from a larger distance, are similar. They are not the same, but the proportions between their parts are the same.

We discussed this in the chapter on proportion. The Seven Sisters mentioned there are not quite similar, but they are similar enough to generate the impression that they are identical, and are only viewed from different distances.

Similarity requires two things: Equal corresponding angles, and equal ratios between edges.

Each condition on its own does not suffice. You need both to hold. Here is an example of two shapes that have the same corresponding angles (all angles are right) but not equal ratios: in the left shape the sides are equal, so the ratio between every pair of edges is 1, on the right hand it is not so.

And here are two shapes having both equal sides, so the proportions of the edges are the same; yet, they have different angles:

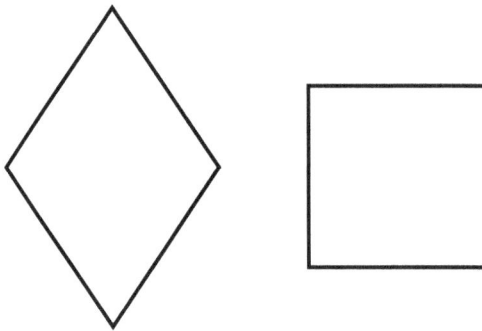

Triangles fare better: any of the two conditions suffices on its own. Two triangles with equal corresponding angles are similar, namely, they have the same ratios between the lengths of their sides; and two triangles with equal corresponding ratios are similar, namely, they have the same angles.

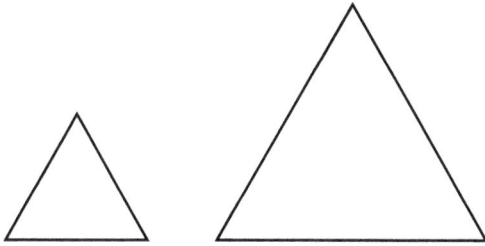

Two equilateral triangles are similar: they have the same angles and the same ratio (namely 1) between the lengths of sides. This shows that the equality of corresponding angles does not suffice for congruence.

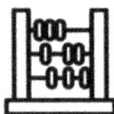

Auxiliary Constructions

One of the best-known theorems in geometry, already mentioned above, is that the sum of the angles in a triangle is 180°, a straight angle. Its proof is a gem — again involving an unexpected element, that was not part of the original problem. An "auxiliary construction."

Choose one edge of the triangle, and call it a "base." Then draw a parallel line to the base through the opposite vertex, like this:

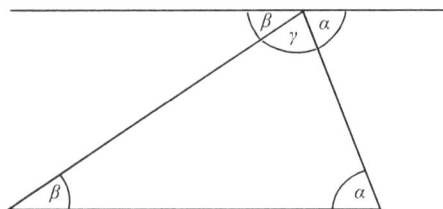

This line is the new, unexpected, element. The justification for denoting two angles in the drawing by α is that they are alternate angles, and hence equal. The same goes for β. The sum of the angles in the triangle is $\alpha + \beta + \gamma$. At the top point, you see that they sum up to a straight angle, namely 180°.

Here is another example of an auxiliary construction. A *parallelogram* is a quadrilateral in which every edge is parallel to the edge opposite it.

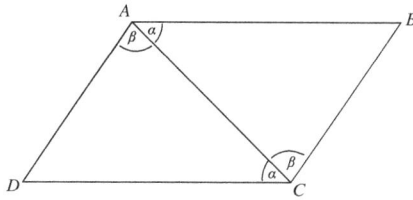

Theorem: In a parallelogram opposite edges have the same length.

The auxiliary construction in this case is natural: one of the diagonals. The pair of angles denoted by α are equal because they are alternate, and the same goes for the pair denoted by β. The two triangles ABC and CAD (in this order!) are congruent by the A-S-A theorem. In this congruence, BC corresponds to AD, and hence they are equal.

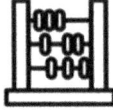

External Angles

In the picture below α is called an "external angle" of the triangle ABC.

Theorem: An external angle in a triangle is equal to the sum of the two angles of the triangle that do not touch it.
    In the picture, $\alpha = \beta + \gamma$.

Proof: $(\beta + \gamma) + \delta = 180°$ (sum of angles in a triangle), and $\alpha + \delta = 180°$. Can you complete the proof?

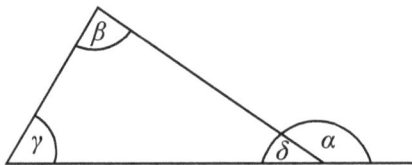

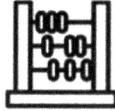

Hierarchy and Converse Theorems

The class of humans is contained in the class of apes, that in turn is contained in the class of mammals (what characterizes them?), that is contained in the class of animals. This means that if you are a human, you are an ape, and if you are an ape, you are a mammal.

"Hierarchy" should be taught starting at grades 3 or 4. Can you find a wider class for "chairs"? And wider than "furnitures"? (I guess "man-made things"). Every chair is a piece of furniture. Is every piece of furniture a chair?

Geometry is an excellent place to learn hierarchy. For example, every equilateral triangle is isosceles, but not the opposite.

Every square is a rectangle. Every rectangle is a parallelogram, but not the opposite. Every square is a rhombus (a quadrilateral with all sides equal). Is every rhombus a square? Every rhombus is a parallelogram, can you draw a parallelogram that is not a rhombus? Use the word "rhombi" — It is an opportunity to teach that in Latin plurals are denoted by an "i" at the end.

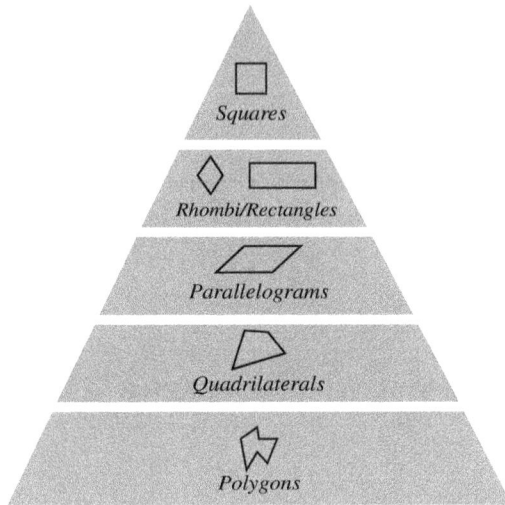

This is a good opportunity to discuss converse theorems.

Every human being is an animal. Is every animal a human being?

Every rectangle is a parallelogram. Is every parallelogram a rectangle? Of course not. Can you show an example?

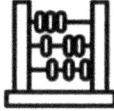

Symmetry

A game is played between two players. They are sitting at a round table, with a round hole in its middle. The size of the hole doesn't matter. Each has a bunch of equally sized coins, for example, a quarter. Each player, in their turn, must place a coin, avoiding the hole and also all the coins placed so far. The first that cannot put a coin loses the game. Who can win?

Let us consider an extreme case: the hole is the size of the entire table. There is no place to put a coin, so the first player loses right at the beginning. He can't put a coin. The second player wins. Is this always true, no matter what the size of the hole is?

Yes: wherever the first player puts their coin, the second does the symmetric move, with respect to the middle. The place is empty since the table is kept symmetric throughout the game, so if there is a place in one spot there is also a place at the symmetric spot.

What happens if there is no hole in the middle? Then the first player wins: they put a coin in the middle, which serves as a "hole." Now he plays by the "symmetric" strategy — whenever player 2 puts a coin, player 1 puts a coin in the symmetric (with respect to the center) position.

The symmetry here was with respect to a point. If O is a point, and P is another point, the symmetric point to P is at the same distance from O as P, in the opposite direction.

P and P' are symmetric with respect to the point O

Such symmetry is called "central." For example, a circle is centrally symmetric (of course it is). But there are more complicated examples. For example: An S is centrally symmetric with respect to its middle point.

There is another type of symmetry — with respect to a line. A shape is called symmetric with respect to a line if using the line as a mirror and reflecting with respect to it preserves the shape.

Question: Is the S shape symmetric with respect to the vertical line going through its middle?

Answer: No, the top right point, when reflected, does not belong to the shape.

Are human beings symmetric with respect to the line going through their middle? So it seems, but not quite. No face and no body are perfectly symmetric. For example, the muscles in the dominant hand are more developed. And inside — well, not symmetric at all. The heart is shifted left, the liver shifted right.

Part 15

The Ruler and the Compass

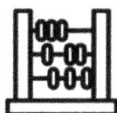

Another Stroke of Genius of the Greeks

Friedrich Gauss, the great German mathematician whose name has already come up above, ordered a drawing of a 17-sided regular polygon to be etched on his tomb. "Regular" means that all sides are equal, and all angles are equal. For example, a square is a 4-sided regular polygon. Gauss was particularly proud of a discovery he had made at age eighteen: that a regular polygon on 17 points can be drawn using solely an unmarked ruler and a compass. His construction answered a problem that was open for over 2000 years. An aftermath remark — the 17-gon was never drawn on his tomb. Göttingen's municipality (where Gauss was buried) accepted the claim of the mason, that having so many sides, the polygon wouldn't be distinguishable from a circle.

But what is this all about, "constructing by ruler and compass"? An example may be clearer than an explanation.

Use a compass and a ruler to find the middle between two points.

$$A \bullet \qquad \bullet B$$

Here is the construction:

- Use the ruler to draw a line L between the two points.
- Use the compass to draw two circles, centered at the two points, having a radius equal to the distance between them. The two circles meet at

two points, one above and one below the line. Connect the two points by a line (again using the ruler). The point C of meeting between the two lines we drew is the middle between the two points.

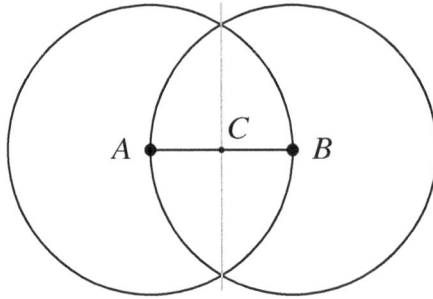

The proof that it works should be left to middle school. For the time being, judge with your eyes.

From this example we learned: the ruler is used to connect points by lines, and the compass to draw points at a certain distance from a given point. Note: we cannot measure lengths. The ruler is not calibrated, it has no distance marks.

For the last two and a half millennia, the ruler and the compass have been the tools used for geometric constructions — no new tools have emerged, an indication that these are indeed natural tools.

On our way, we constructed a right angle: the two lines we drew are perpendicular. So, with a ruler and a compass you can draw a right angle.

Next,

Construct a perpendicular to L that goes through a given point P.

Put the apex of the compass on P, and mark two points A and B on L equidistant from P. The perpendicular bisector between A and B, which we learned how to draw, is perpendicular to L and passes through P.

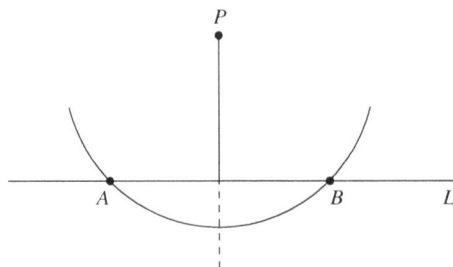

Drawing a parallel to a given line *L*, going through a given point *P*.

Remember our friends from Euclid's "fifth axiom"? The fifth axiom was about the existence of a parallel line through the point *P*. So, we know it exists. Now we construct it.

Use the compass to draw a circle with center *P* that meets *L* (see the left part of the drawing below). Call its radius "*R*." Let it meet *L* at points *A* and *B*. Use the compass to make a copy of *AB* on *L*, call this new segment *A'B'* (see picture below). Draw two circles with radius *AP*, one centered at *A'* and the other at *B'*. They meet at a point *P'*. The line *PP'* is parallel to *L*, because *P* and *P'* have the same distance from *L*.

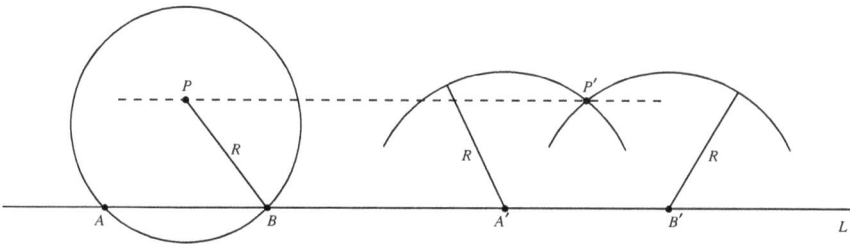

Replicating an angle. An angle α is drawn, with apex *A* and rays *L* and *M* and a point *A'* outside it is given. Draw an angle equal to α and having *P'* as an apex.

Solution: Draw through *A'* a line *L'* parallel to *L*, as we just did in the previous section, and a line *M'* parallel to *M*.

How do we know that *L'* and *M'* form an angle α? This is part of the definition of "equality of angles." Two angles formed by corresponding parallel lines are equal. There is no need for further proof.

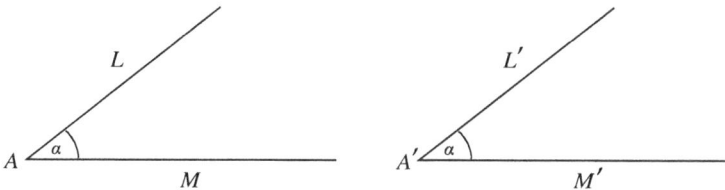

Impossible Constructions

The ancient Greeks left three major unsolved construction problems, all three seemingly simple.

Dividing an angle into three equal parts ("trisection" of the angle).

Doubling the cube: Given the side of a cube, draw, using a compass and ruler, the side of a cube whose volume is twice that of the original cube.

The third and most famous of the three is as follows:

Squaring the circle: Given a circle, construct the side of a square having the same area. (One can ask also for a square having the same circumference.)

Two millennia of struggle produced only frustration, until in 1837 Pierre Wantzel, a French mathematician, proved that the first two were impossible. The third became a synonym for an impossibly difficult task even before its impossibility was rigorously proved. This was done in 1882, by the German mathematician Lindemann.

Another, slightly less known construction task was

Constructing a heptagon, a regular polygon with 7 sides. Two thousand years of struggle got geometers nowhere until young Gauss proved that it was an impossible task.

The proofs of all four impossibilities use advanced algebra and are beyond the scope of this book.

Part 16

Length, Area, and Volume

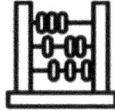

Units of Length

Ask your students to draw a segment of length 3.

This is like "give me 3" — 3 what? There must be a denomination. Start a discussion — the students will probably suggest "3 centimeters," or "3 meters," or "3 inches."

For numbers, Nature itself offers the units: 3 flowers, or 3 stones. For measurements, we must invent a unit: a meter, a foot, a versta (a Russian unit of measure, about one kilometer). The choice is arbitrary: we just decide what a "meter" is.

Do not hesitate to use the word "arbitrary" with the children, it is an important notion. You realize the arbitrariness by the fact that different countries chose different measure units. The Anglo-Saxon world chose inches and feet (= 12 inches) and yards (= 3 feet).

The meter was defined by the French in 1791 (after the French Revolution) in such a way that the distance from Paris to the North Pole is ten million meters. There was some logic to it — you get a measure that is close to the English yard and unlike the "yard" fits in the decimal system. It is a useful measure — things around us measure a few meters, so it is human-friendly. To measure smaller things, we divide by 100 to get the centimeter. An opportunity to ask the origin of the word — do you know any other "cents"? Of course, 100 cents in a dollar, centipede, percent. What will be a centiliter? For larger measures we coin the term "kilometer," and smaller — millimeter.

How many millimeters are in a kilometer?

The children enjoy the fact that there is something around them that is counted by the millions (this is the answer to the question). Another, more surprising fact: a cubic meter contains a million cubic centimeters. And a cubic centimeter is sizable. We do not need a microscope to see it.

When you speak about large numbers, tell them about the number googol, 1 with 100 zeros (the search engine Google borrowed its name from it.). I once asked my daughter whether there is anything in the world of which there is a googol. "Yes," she said, "the googolth part of a second" (just as there are one hundred hundredths of a second in each second).

Square Centimeter and the Area of a Rectangle

We also need a unit for measuring area. A square with side one centimeter is one such choice.

It is called a "square centimeter."

A rectangle with side lengths a, b centimeters has an area $a \times b$ square centimeters. This is seen in the following picture:

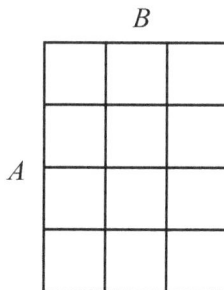

The width is 3 centimeters, the length is 4, so there are 3×4 squares. Each has an area of 1 (1 square cm) since each is a 1cm × 1cm square.

233

What happens if we stretch the rectangle horizontally 3 times? Its area is multiplied by 3. And if we stretch it vertically 3 times? The same — the area grows by a factor of 3. And if we stretch 3 times in both directions, that is, if the entire rectangle is multiplied by 3? The area is multiplied by 3×3, namely by 9. In general, if we multiply a rectangle by x, its area is multiplied by x^2.

Since any shape can be covered by rectangles, at least to a good approximation, we have:

If both dimensions of a shape are multiplied by a number x, its area is multiplied by x^2.

In the next chapter we shall use this fact to prove the famous Pythagoras theorem.

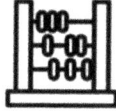

Pythagoras' Theorem

Pythagoras' theorem has an attractive name and an intriguing history — Pythagoras himself was a colorful figure, who led a mathematical cult. But these are not the only reasons for its acclaim. It acquired its fame mainly because of its usefulness: it enables calculating distances.

The theorem provides a connection between the lengths of the sides of a right-angle triangle.

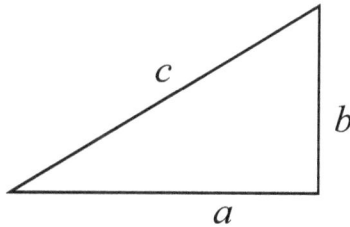

$$c^2 = a^2 + b^2$$

Or, in a drawing:

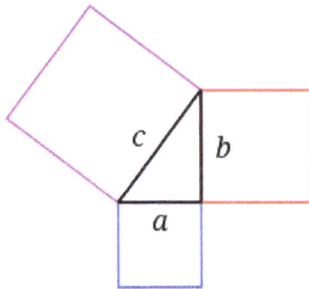

The area of the square built on the hypotenuse of length c is c^2, and the theorem says that it is equal to the sum of the areas of the squares on the two perpendicular edges. $c^2 = a^2 + b^2$. So, if you know a and b you know c. For example, if $a = 3$ and $b = 4$, then $c^2 = 3^2 + 4^2 = 9 + 16 = 25$. Namely, $c = 5$. Caution: In general, the result is not an integer.

This means that if we know how to calculate east-west distances (a in the picture) and north-south distances (b in the picture), we can calculate any distance (c in the picture).

As to proofs, there are hundreds. But the arguably most illuminating one starts from the picture below. Note: AD is perpendicular to BC:

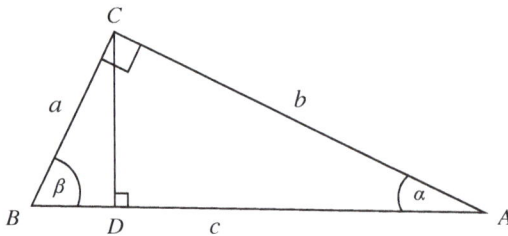

There are three triangles in this picture: BAC (the large one), CAD, and BCD. As we shall see, Pythagoras' theorem is tantamount to the obvious fact that the area of the big triangle is the sum of the areas of the small triangles! The crucial observation is that **all three are similar** — they have the same angles. Here is why BCD is similar to BAC: both have a right angle, and they both share the angle β. The equality of the third angle comes for free: since the sum of the angles in any triangle is 180°, if two triangles share two angles, they share also the third.

The same goes for the similarity between CAD and BAC.

The similarity of *BCD* and *BAC* means that the first is obtained by multiplying the second by some fixed number x. What is x? It is the ratio between the corresponding sides. We know the ratio between the hypothenuses: the hypothenuse of *ABC* is c, and the hypothenuse of *BCD* is a. So, the ratio is $x = \frac{a}{c}$.

Denote the area of *ABC* by S. By what we learned in the previous chapter, the area of *BCD* is $x^2 S = \left(\frac{a}{c}\right)^2 S$.

Similarly, the area of *ACD* is $\left(\frac{b}{c}\right)^2 S$.

Now comes our observation — the sum of the areas of *ADB* and *ACB* is S — the two triangles partition the triangle *ABC*.

This means:

$$\left(\frac{a}{c}\right)^2 S + \left(\frac{b}{c}\right)^2 S = S$$

Canceling out S, we get

$$\left(\frac{a}{c}\right)^2 + \left(\frac{b}{c}\right)^2 = 1, \quad \text{meaning:} \quad \frac{a^2}{c^2} + \frac{b^2}{c^2} = 1$$

There is a common denominator, so, the equality is $\frac{a^2+b^2}{c^2} = 1$

The value of a fraction being 1 means that its numerator is equal to its denominator. Hence $a^2 + b^2 = c^2$.

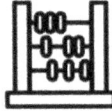

The Area of the Triangle and the Volume of the Pyramid

We culminate the part on geometry with the area of the circle, and the volume of the ball.

This will necessitate calculating the area of the triangle and its three-dimensional analog, the volume of the pyramid.

To calculate the area of the triangle, look at the following picture:

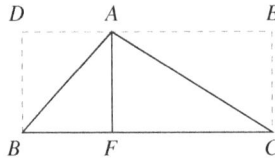

What is the area of the triangle ABC?

The area of the triangle AFC is half the area of the rectangle $AECF$, and the area of the triangle AFB is half of the area of the rectangle $ADBF$. Summing, the area of ABC is half of the area of the rectangle $DECB$. So, the area of the triangle is half of the base times the height.

$$\text{Area of a triangle} = \frac{1}{2}(\text{base} \times \text{height}).$$

Next, an ambitious leap: the volume of the pyramid.

What do we mean by "pyramid"? The Egyptian pyramids have a square base, and their apex (top) is connected by segments to all points in the base — the pyramid consists of all the points on all these segments.

The mathematical term "pyramid" is more general. The base does not have to be square. It can be any shape. If the base is a triangle, then the pyramid is called "triangular." If it is a square, we call it a "square pyramid." And if it is a circle — no, we don't call it a "circular pyramid," but a "cone."

Do you remember what the height of a triangle is, to a given base? It is the length of the segment perpendicular to the base, from the apex. Here, again, the height is the length of the shortest segment from the apex to the base, and we know that "shortest" means it is perpendicular to the base.

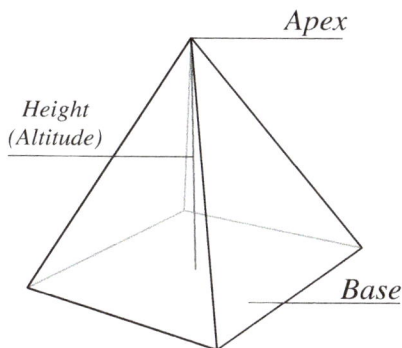

Apex

*Height
(Altitude)*

Base

Here is the formula:

$$\text{Volume of a pyramid} = \text{area of base} \times \text{height} \times \frac{1}{3}.$$

This is similar to the formula for the area of a triangle — remember? There it was the length of base, times height, divided by 2.

Can you guess where the 2 and 3 come from? Yes, you are right — this is the dimension. A triangle lives in 2 dimensions, and the pyramid in 3. Can you guess the formula in 4 dimensions? Yes, there is a quarter in the formula. But it is hard to visualize "volume" in four dimensions.

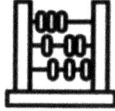

One Example Suffices!

How do you prove the formula for the volume of the pyramid?

Let me show you an example in which the formula is valid. Take a cube with all sides of length 1. Its volume is $1 \times 1 \times 1 = 1$. Connect its middle with all sides. This partitions the cube into 6 square pyramids — why 6? Because a cube has 6 faces, as every dice-tosser knows.

The volume of each pyramid is then $\frac{1}{6}$ of the volume of the cube, namely $\frac{1}{6}$. The base has an area $1 \times 1 = 1$. The height is $\frac{1}{2}$, half the length of the cube, because the apex is in the middle. According to our formula, the volume of each pyramid is

$$\text{area of base} \times \text{height} \times \frac{1}{3} = 1 \times \frac{1}{2} \times \frac{1}{3} = \frac{1}{6}$$

Just right. In this example, at least, the formula works. We have calculated the volume of one particular pyramid. Let us call it the "model pyramid." What about general pyramids? Here we are in for a surprise — it is enough to prove the formula for one particular pyramid, the model! A single case will prove it all! The other cases will follow suit! How? I will explain it in the next chapter. It is a bit abstract — you are allowed to skip.

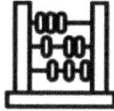

The Cavalieri Principle

At the core of the calculation of the volume of pyramids stands the following claim:

Any two pyramids with the same height and same base area have the same volume.

The proof of this fact is done by a method that was already known to Archimedes and was formulated explicitly by the Italian mathematician Cavalieri. It is a precursor of integral calculus, in which the whole is partitioned into minute parts, to be later integrated. It is done by slicing the two pyramids, as in the following picture.

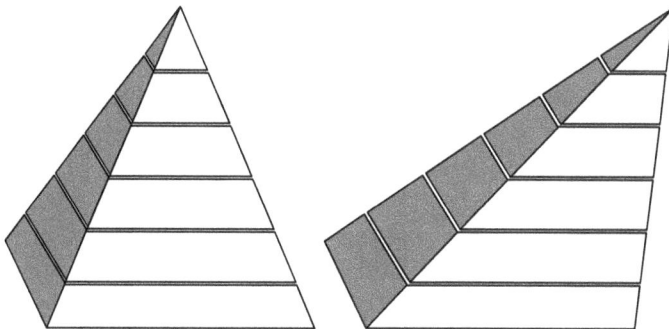

We are assuming that the two pyramids have the same base areas and the same height. The crucial observation is that each slice on the left has the same volume as the corresponding (same height above ground) slice on the right. Obviously, the two slices have the same thickness — that's what slicing is about. But they also have the same base area. Why? Take for example the middle slices in the two pyramids. By triangle similarity, they are each half of the base, lengths-wise. So by a rule we learned, the area of each of the middle slices of the pyramids is a $\left(\frac{1}{2}\right)^2 = \frac{1}{4}$ of the area of the base. Since the areas of the bases of the two pyramids are assumed to be the same, it follows that so are the areas of the middle slices. A similar argument goes for all slices. For example, the slice a third of the way from the top has $\frac{1}{9}$ of the area of the bases, in both pyramids.

If we make the slices thin, the volume of each slice is very close to its base area times its height (thickness). So, the two corresponding slices (that have the same thickness and same area) have ("in the limit," namely when the thickness of the slices tends to zero) the same volume. The volumes of the two pyramids are the sums of the volumes of the slices; so the two pyramids have the same volume.

What does it mean "in the limit"? The Greeks, who invented this method, never bothered to define this notion rigorously. Neither did Cavalieri, nor Newton and his contemporaries, who used the method. A totally rigorous proof had to wait until the 19th century. In this book we are still in elementary school, so we allow following the intuition of the Greeks.

So far, we have proved the formula for any pyramid having the same base area and height as the "model" pyramid, regardless of shape. What do you do with a pyramid with a general height? Stretch it or shorten it so that the height is the same as that of the model pyramid (the height of that one, you may remember, is $\frac{1}{2}$). If you stretch it, say, 5 times, the volume will grow by a factor of 5. If the formula is true after the stretching, it is true also before, because both sides were multiplied by 5 (in the example), so if equality holds now, it held also before. The same argument goes for the area of the base — given two pyramids, stretching (or compressing) will make the areas of their bases equal.

Summarizing: every pyramid can be stretched so as to make its height 1/2, the height of the model pyramid, and its base area 1, the area of the base of the model pyramid. It remains to observe that since the formula for the volume is valid after the stretching (we know that), the formula was valid also before the stretching, namely in the pyramid we started with.

Part 17

The Magic Circle

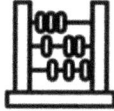

Geometric Locus

"Locus" means "place" in Latin. Students may know the words "local" and "location."

Can you describe the locus of Mary in the class? She sits in the second row, third from left ("left" for the students, for the teacher it is right.) John, can you describe the locus of your home? 314 Pythagoras street? Good, what floor?

In these examples the description is unambiguous. You know precisely where to look for Mary and for John. It becomes more interesting when the description is less specific. Like "I live in London."

Here is an example of ambiguity. A driver goes from town A to town B, 100 kilometers away. His car breaks down and he calls his wife and tells her "I am 10 kilometers from the middle between A and B." Can she know for sure where he is? Not quite. There are two possibilities: 10 km before the middle and 10 km after the middle.

Exercise: Another driver on the same route says "I am closer to A than to B," what can you say? (Answer: he has not yet reached the middle, he is somewhere between A and the middle).

Our next hero is a polar adventurer, who calls his rescuers and tells them — "I am 50 kilometers from the North Pole" — do they know where he is?

Answer: He is on a circle of radius 50 km, whose center is at the North Pole.

A circle is a "geometric locus":

A circle of radius R and center O is the locus of all points that have a distance R from O.

Another traveler calls his rescuers and tells them — "My distance from the pole is 50 km or less" — where is he? Right — inside the circle.

What is the locus of all points inside the classroom that are of a distance of 1 meter from the wall of the board?

Discuss with the students: how many data items are needed to specify a straight line? Of course, two points are enough. Is one point on the line enough? A point on the line and another line which is perpendicular to it also suffices. So, we need two pieces of data.

How about 3 points on the line? Too much information: most likely there is no line passing through all three points.

To define a line, you need two pieces of data.

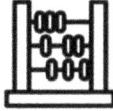

The Perpendicular Bisector

What are the points that are equidistant from New York and Boston? More generally:

Given two points, *A* and *B*, what is the locus of the points whose distance to *A* is equal to their distance to *B*?

Ask your students: can you find one point whose distance to *A* is equal to its distance to *B*? The students will probably point at the middle of the segment connecting *A* and *B*. Correct. Can you find another? And another? And another? With some patience, they will draw lots of points, all on one line. Here it is:

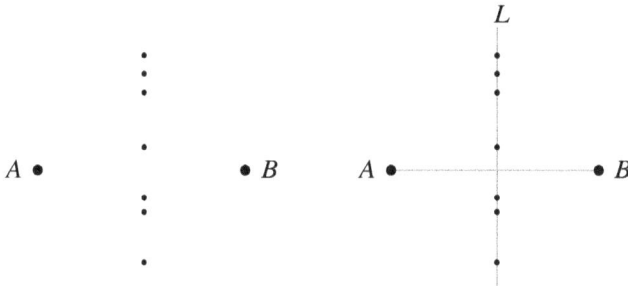

On the left are some points that are equidistant from *A* and from *B*. On the right: the line containing all these points.

This line is called the "perpendicular bisector" of the segment joining *A* and *B*. It is the line perpendicular to the segment *AB* and bisecting it.

We have guessed that:

The geometric locus of the points equidistant to *A* and to *B* is the perpendicular bisector of the segment *AB*.

Note: to prove this you have to show both directions. That points on the perpendicular bisector are equidistant from *A* and *B*, and that every point equidistant from *A* and from *B* is on the perpendicular bisector. We will skip the proof, and just hint that the proofs use congruence of triangles.

Here is a nice application of the notion of "geometric locus." Try to get the students to observe:

Theorem: The perpendicular bisectors to the three sides of a triangle meet at one point.

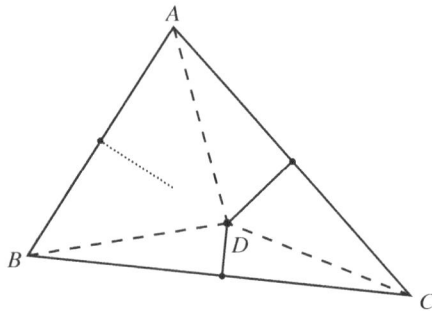

Proof: Let's call the triangle *ABC*. Let *D* be the meeting point of the perpendicular bisector of *AC* with the perpendicular bisector of *BC* (they do meet, they are not parallel — why?). We must show that *D* is also on the third perpendicular bisector, that of *AB*.

Since *D* is on the perpendicular bisector of *AC*, *DA* = *DC*.
Since *D* is on the perpendicular bisector of *BC*, *DB* = *DC*.
So, *DA* = *DB*. Namely, *D* is equidistant from *A* and *B*.

Remember? All points equidistant from A and B are on the perpendicular bisector of segment AB. This is what a "geometric locus" is. So, D is on the perpendicular bisector of AB.

So, D is on all three perpendicular bisectors.

So, all three perpendicular bisectors share the same point, D. As promised.

(In mathematics, "As promised" is written "QED" — quod erat demonstrandum — what had to be proved.)

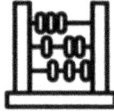

Angles in the Circle

As a child, I was highly impressed when I learned that the two angles in the picture, both leaning on the diameter of the circle, are equal — both are straight, namely 90°.

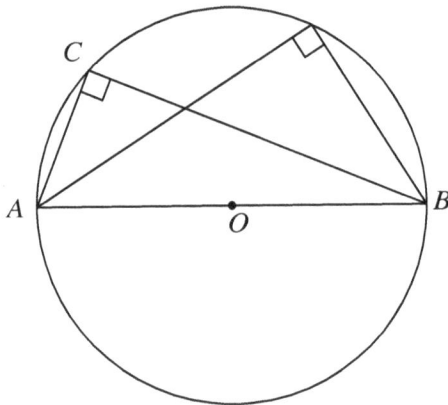

Another way to put it is that all points on the circle "see" the diameter at an angle of 90°. Let us show why any point C "sees" AB with an aperture of 90°.

The proof starts with the most natural auxiliary construction: connecting C to the center, O. This is natural, because you know something about it — its length is the radius.

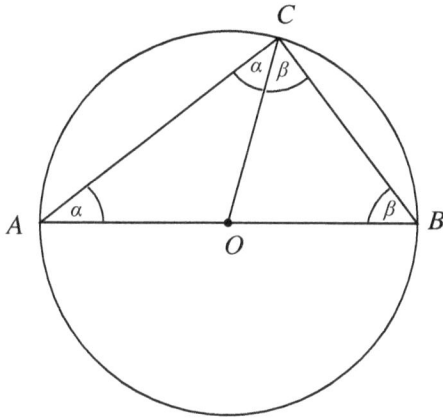

The triangle *ABC* is now partitioned into two triangles: *OAC* and *OBC*. Both are isosceles, namely have two equal sides, because two of the sides are radii of the circle. Therefore, they have equal base angles — α in *OAC*, and β in *OBC*. The sum of the angles in the triangle *ABC* (which is, of course, 180°) is then $2\alpha + 2\beta = 2(\alpha + \beta)$. So, $2(\alpha + \beta) = 180°$, so $\alpha + \beta = 90°$.

Look at the picture to see that $\alpha + \beta$ is the angle we are speaking about — the angle by which *C* sees the diameter.

What happens if we consider the angle seeing *any* segment, not necessarily the diameter? Here are two angles seeing the same arc *AB*. Can you guess the theorem?

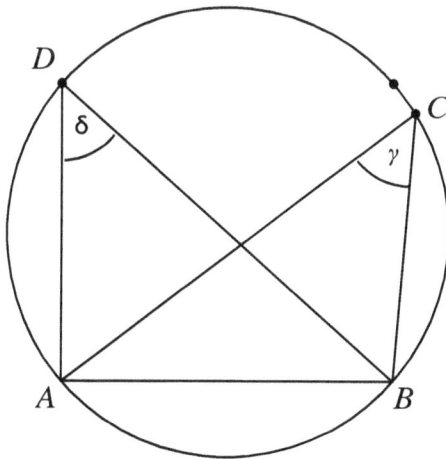

Yes, you are right: the two angles are equal. $\delta = \gamma$. All angles seeing the same arc are equal. The secret:

Fact F: A circumference angle is half of the central angle on the same arc.

Why is this true? Look at the following drawing:

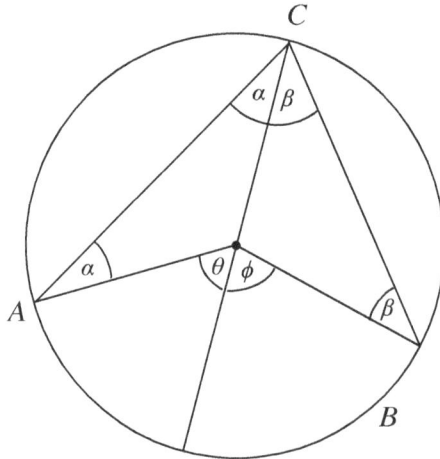

θ and ϕ are external angles, each in its triangle, and hence $\theta = 2\alpha$ and $\phi = 2\beta$. So

$$\theta + \phi = 2(\alpha + \beta).$$

Look at the picture and see: $\theta + \phi$ is the central angle, while $= \alpha + \beta$ is the circumference angle — we have reached our goal.

What happens if AB is a diameter? Then the central angle is 180°, so the circumference angle is half of it, 90°, as we already know. A circumference angle on the diameter is a right angle.

$$\pi$$

You have probably heard about the mysterious number Pi. It has a central role in geometry, but surprisingly also in number theory. Many also know that its value is approximately 3.14. Not precisely, and no finite number will pin it down. Johan Heinrich Lambert (1728–1777)[1] proved that π is not a fraction — so its decimal representation is infinite:

$$\pi = 3.14159\ 26535\ 89793\ 23846\dots$$

It goes on and on, never to end, and is not even periodic, namely, it does not have a repeating pattern (it cannot be, for example, 3.141414...).

But what is it? π, the Greek letter "p", stands for "perimeter." It is the ratio between the perimeter (circumference) of a circle and its diameter. "Diameter" is the distance between two opposing ("antipodal," is the formal name) points on the circle. It is 2 times the radius, as can be seen from this picture:

---

[1]A curiosity: Newton's birth year, 1642, coincided with Galilei's death year. Gauss' birth year coincided with Lambert's death year.

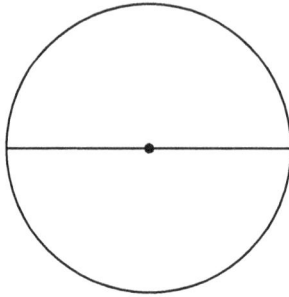

So, π is approximately 3.14. Can we prove that it is larger than 3? Let us first be less ambitious and show that it is greater than 2.

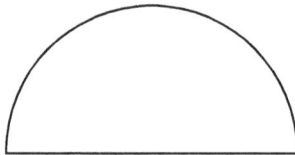

Here is a drawing of half a circle, with the diameter "closing" it. What is longer, the half-circle, or the diameter? Of course, the half circle. The diameter is straight, and the straight line is the shortest path between its endpoints. The whole circle is two halves, and since each is larger than the diameter, the perimeter is at least 2 times the diameter. The ratio between the two is at least 2, so $\pi > 2$.

To see that π is larger than 3, look at the following picture, where the angles at the center are drawn so that they are all equal:

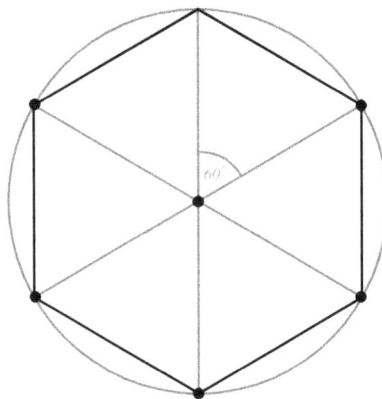

Fact: All segments in the picture are equal, they are all R, the radius of the circle.

This means that the triangles in the picture are equilateral — all their sides are equal. This follows from the fact that all their angles are 60°.

The circumference of the hexagon is $6R$, and the circumference of the circle is still larger — so it is at least $6R$. Since D, the diameter, is $2R$, we have $6R = 3D$. Namely, the circumference C of the circle satisfies

$$C > 6R = 3D.$$

So π, the ratio of the circumference to the diameter, namely C/D, is greater than 3.

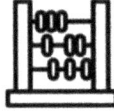

The Area of the Circle:
Is Earth Flat After All?

In the summer of 1666, twenty-four-year-old Isaac Newton returned to his native village, Woolsthorpe, to escape the plague that rampaged in London. During that summer he invented Calculus (I will promptly explain what it is,) discovered the law of gravity, and set the basis for modern optics. "I was standing on the shoulders of giants," he explained the miracle. No, he was not a poet on top of being a great mathematician — this was a common saying in his time. And no, he was not humble — he bickered for years with the German mathematician-philosopher Leibnitz on the seniority in the invention of calculus. And yes, he was standing on giants' shoulders, in particular one, who developed the rudiments of calculus 2000 years earlier — Archimedes.

Why did the ancients (and some present-day cranks) believe Earth is flat? Because they observed it from close. They did not have satellites, from whose point of view Earth is a ball. An ant, walking on a sphere, believes she is on a plane. From close, every smooth surface looks flat. This approach is what stands at the core of one of Archimedes' nicest achievements — the calculation of the area of the circle.

Before explaining how he did it, let me tell you what he proved. The area of a circle of radius R is:

$$Area = \pi R^2$$

For example, the area of a circle of radius 10 cm is $\pi \times 100$, around 314 square cm. In the formula, the radius is raised to the power of 2, since we are in 2 dimensions. By the formula, if R is multiplied by (say) 3, the area is multiplied by $3^2 = 9$, a fact we have already noticed to be true for any 2-dimensional shape.

For the proof, look at this picture:

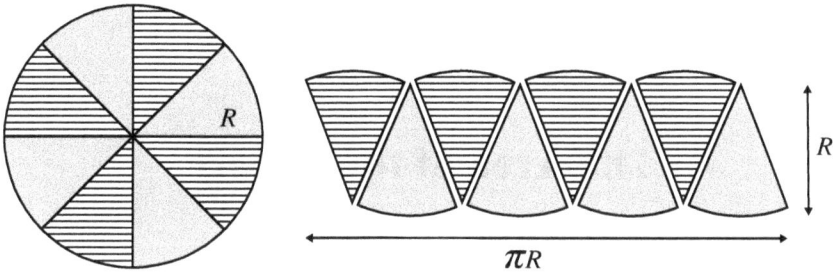

We (or rather Archimedes) divided the circle into many sections (the formal name for "pizza slices"). We choose the sections to be so small, that their "base" (the part on the circle) is almost straight. So, each section is almost a triangle. The height of the triangle (distance of the center of the circle to the base) is R, the radius (see the picture). The area of this triangle is $\frac{1}{2}$ times the base times the height, namely $\frac{1}{2}BR$. So, the area of the circle, which is the sum of all areas of the triangles, is $\frac{1}{2}R$ times the sum of all bases (we used distribution here). But the sum of all bases is the perimeter of the circle, which is $2\pi R$. So, the area of the circle is $2\pi R \times \frac{1}{2} \times R$, the two 2s cancel out and we remain with $\pi R \times R = \pi R^2$.

The picture on the right suggests a clever way — arranging the pizza slices as in the picture you get a shape close to a rectangle, with a width half of the circumference (the circumference is the sum of the top and the bottom lines) and height R, so its area is πR^2.

This was an approximation because the slices are not really triangles. But "in the limit," it is precise — the smaller the slices are, the closer they are to be triangles.

What we calculated here is called an "integral." An integral is the sum of many small (smaller and smaller, more and more) parts. It is denoted by $\int$, to remind us of "S," the first letter in "Sum."

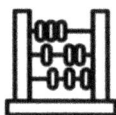

The Area of the Sphere

Abe, Bob, Carol, Diane, and Ebenezer are three brothers and two sisters. Their favorite uncle brought them a ball of chocolate. The uncle is health aware, and hence the ball was empty — a hollow sphere, just the shell. The question was how to divide it among the children. Their mother had an idea — she put the sphere through a bread-slicer and sliced it into 5 rings of equal width (a very delicate bread-slicer it was, no ring broke).

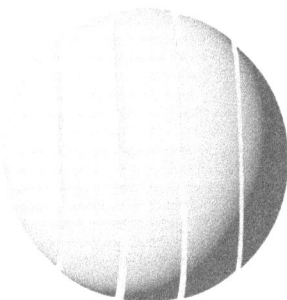

Now she let them choose. Of course, each of them wanted to maximize the amount of chocolate in their piece. Abe and Ebenezer said — "We shall take the extreme pieces — they do not have holes in them." Carol chose the middle one because it has the largest radius. Bob and Dian chose the rings next to the extreme ones — they do not have the largest radius, but they are slanted, so there is more chocolate in them.

Who was right? If you have any experience with mathematicians, you may have guessed: somebody is pulling your leg. The surprising answer is that they are all the same. The five rings have the same area (and hence the same amount of chocolate — remember that this is a shell). The slantness of the side rings precisely compensates for the small radius. This is not hard to prove, using similarity of triangles, but let me not tax the reader's patience too much.

This is the crux of the proof of another of Archimedes' discoveries:

The area of the sphere of radius R is $4\pi R^2$.

The radius is squared because we are speaking of area, and area is 2-dimensional.

To realize this, look at the middle ring. Its width is a fifth of the width (diameter) of the sphere (the bread slicer sliced into 5 equal parts), which is $\frac{2R}{5}$ (the diameter is $2R$).

$$2R\,/\,5$$

$$2R$$

It is a ring, there is no slantedness (assuming we divide it into many slices, each looks straight, in particular, the middle). Mathematically, it is a cylinder. It is easy to convince yourselves that its area is its circumference (which is $2\pi R$) times the width, which we found to be $\frac{2R}{5}$ (unwrapping it produces a rectangle, whose area is the product of its two sides). So, the area of the middle slice is $\frac{2R}{5} \times 2\pi R$, which is $\frac{4\pi R^2}{5}$. If you believe me that all 5 slices have the same area, the total area of the sphere is 5 times this area, namely $5 \times \frac{4\pi R^2}{5}$, which is $4\pi R^2$.

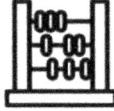

The Volume of the Ball

The volume of the ball with radius R is:

$$\frac{4}{3} \pi R^3.$$

Archimedes, who discovered this formula, regarded it as the pinnacle of his achievements, so much so that he asked that a drawing expressing it would be etched on his tomb (nobody knows whether this wish was fulfilled).

This is a very natural formula: the R^3 is there because we are in 3 dimensions. π should be there, no doubt. So, the question is only why $\frac{4}{3}$.

You may have guessed: the $\frac{1}{3}$ comes from the formula for the volume of a pyramid. Archimedes put a point in the middle of the ball and partitioned the ball into many (the more the merrier, namely the better the approximation to the truth) pyramid-like parts whose apex is the center.

Since there are many (many!) parts, with very small bases, we may consider the bases to be flat, namely that each part is indeed a pyramid, not only "pyramid-like." As we said, from close the earth looks flat. Hence we can apply the formula for the volume of the pyramids. Namely,

The volume of each small pyramid is R (the height) times the area of its base, divided by 3.

The volume of the ball is the sum of the volumes of the small pyramids, so it is the sum of the areas of their bases, times R, divided by 3. Since the bases cover the entire area of the sphere, the sum of the areas of the bases is just the area of the sphere, which we found to be $4\pi R^2$. So, the volume is $\frac{1}{3} \times R \times 4\pi R^2$, which is $\frac{4}{3}\pi R^3$, as promised.

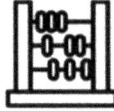

Beyond the Firmament:
Middle School Mathematics

A monk peeping beyond the edge of the universe

This is a famous picture, called "Flammarion," of unclear origin.[2] I would like to do what the monk is doing in the picture — look beyond. What do the students learn in middle school?

---

[2] It first appeared in a book in 1888, but the style is Renaissance.

Around the age of 12, there is a leap in the child's capacity for abstraction. For example, they can understand a non-tangible notion such as negative numbers. It needs imagination to understand that "owing 100 dollars" is "having minus 100 dollars." Negative numbers are one of the first topics in middle school.

But the main leap of abstraction is algebra. The invention of the number requires abstraction: understanding that it doesn't matter whether these are 3 apples or 3 pencils, the rules of arithmetic will be the same for both. Algebra is the next level of abstraction — talking about *all* numbers. There are rules that are valid for every number, no matter which. To express such rules, we use letters that can denote any number. For example, $x + 1 = 1 + x$. This is true whatever x is. Or the more general $a + b = b + a$ — the commutative law of addition. In this role, the letters are called "variables." Formulas are economical. "Adding 1 to any number increases it" is written concisely as "$x + 1 > x$." So, algebra uses letters to generalize.

A letter can also be used to denote a number that is temporarily general because we don't know it. We search for it, using some information. In such a role it is called an "unknown," and if the information is expressed by an equality sign, it is called an "equation." For example, $2x = 10$ is an equation, whose solution is $x = 5$. Not always does the information determine the number uniquely. The information "his family name is Jones" does not determine a single person. Likewise, the information $(x - 1) \times (x - 2) = 0$ does not determine x. Can you find two different solutions? This is called a "quadratic equation."

There is an entire world waiting.

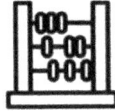

Epilogue: The Infinitude
of Mathematics

The story goes that Brahms and Mahler stood together one day on a bridge overlooking the Danube. "You know," said Brahms to the aspiring composer who was to become his heir, "we are the last."

In response, the young Mahler pointed at the river, and said — "look, there goes the last ripple."

Mathematics will never be exhausted. The Danube will dry long before all mathematical questions are settled. Possibly mathematicians will be replaced by artificial intelligence (AI) — but even AI will not exhaust all. New concepts and new questions will always emerge and re-emerge.

Students in elementary school are exposed only to the rudiments. But "rudimentary" doesn't mean "shallow." The decimal system, the algorithms for calculating the operations, the place-value system, fractions, Euclidean geometry, and even the concept of the "number" — all these were developed by the ancients with great wisdom and effort. The combination of being at the same time elementary and deep makes for beauty. Elementary mathematics is as beautiful as advanced mathematics. Conveying this beauty to the children is the best way to make them love the subject and understand it. There is no real understanding without experiencing the beauty and subtleties.

In the book, I tried to go into the fine mesh that constitutes elementary mathematics and to understand the rationale behind its algorithms, facts, and concepts. I believe that if these sink into the mind of the teacher — be it parent or schoolteacher — they will be passed on to the children.

Index

www.ingramcontent.com/pod-product-compliance
Lightning Source LLC
Chambersburg PA
CBHW050546190326
41458CB00007B/1942